A2 Geography
UNIT 6

Edexcel

Specification **B**

Unit 6: Synoptic Unit
(Issues Analysis)

Sue Warn and David Holmes

Philip Allan Updates
Market Place
Deddington
Oxfordshire
OX15 0SE

tel: 01869 338652
fax: 01869 337590
e-mail: sales@philipallan.co.uk
www.philipallan.co.uk

© Philip Allan Updates 2003

ISBN-13: 978-0-86003-695-1
ISBN-10: 0-86003-695-2

This Guide has been written specifically to support students preparing for the
Edexcel Specification B A2 Geography Unit 6 examination. The content has been
neither approved nor endorsed by Edexcel and remains the sole responsibility
of the author.

Material for the Aqaba issues analysis exercise (pp. 71–81) was collected by
D. Balderstone while on a 'Discover the world' goodwill visit to Jordan.
Information for the Soufrière issues analysis exercise (pp. 82–95) was compiled
by Sue Warn on a Darwin Initiative visit to St Lucia. The map on page 44 is
reproduced by permission of Ordnance Survey, on behalf of The Controller of
Her Majesty's Stationery Office, © Crown Copyright 100027418.

Printed by MPG Books, Bodmin

Contents

Introduction

■ ■ ■

Synoptic Guidance

■ ■ ■

Questions with Answer Guidelines

Introduction

About this guide

The purpose of this guide is to help you understand what is required to do well in **Unit 6: Synoptic Unit (Issues Analysis)**. The guide is divided into three sections.

This **Introduction** explains the process of issues analysis. It also provides some general advice on how to approach the unit test.

The **Synoptic Guidance** section sets out the skills and techniques you might need to use when carrying out issues analysis and outlines how to handle the pre-release resources. Diagrams and examples are given to help your understanding.

The **Questions with Answer Guidelines** section contains two examples of issues analysis, in the style of the unit test. Suggested frameworks for answers precede the resources and answer guidelines.

Issues analysis exercises

Issues analysis exercises combine problem solving with decision making.

The structure of the task

Pre-decision research

This involves:

- assessing the role of key players
- considering conflicts at national, regional and local scales
- analysing the resources to provide a targeted synopsis as a background to an issue (using your geographical knowledge)

Decision making

This means:

- looking at the options and ranking them from best to worst
- assessing environmental and socioeconomic costs/benefits
- comparing strengths and weaknesses of options and matching schemes to sites (synthesis)
- exploring who are 'the players' and what influence they may have on the decision
- exploring aspects of decision making about an issue

Post-decision phase

This involves:

- monitoring the likely impact
- assessing sustainability

- developing fieldwork programmes for this assessment and monitoring
- developing priorities for action
- developing schemes for the future
- looking at the impact of parallel developments
- taking the decision further (looking at further exemplars)

The process

Identify task → Fix objectives → Note supporting evidence and analyse it

Use prior geographical expertise

Evaluate options, impacts, costs and benefits ← Generate alternatives ← Organise information in format required by summarising key features

Target the best option → Make the decision → Explain/justify the choice

Make decisions for the future/prioritise future plans ← Monitor impacts and solve operational problems ← Implement the plan

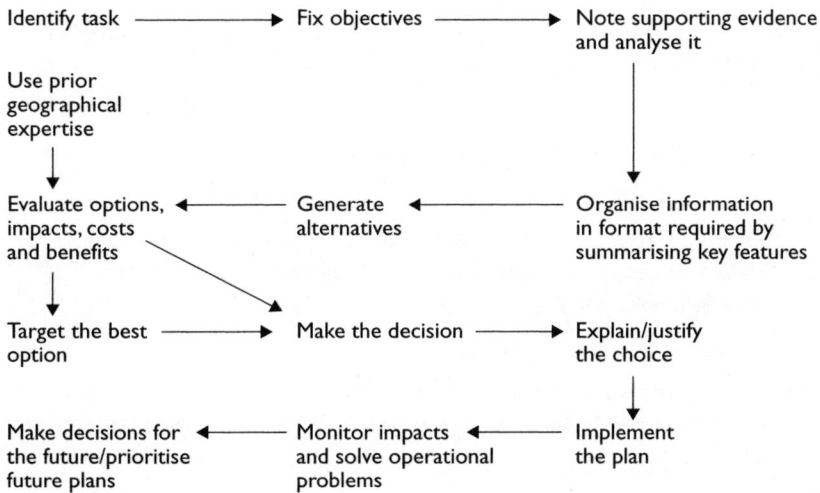

For each issues analysis exercise, there is a huge variety of tasks that could be set, depending on the issue chosen. Expect a series of linked tasks focused on a place or area and environment unfamiliar to you. Examples of scenarios include:
- small or local scale
  - coastal development options
  - designing sustainable options for a city
  - development options for a small island economy
- regional scale
  - river management schemes
  - managing a National Park
  - a new industrial complex for a region
- national scale
  - award of World Heritage status or cultural city status
  - airport development options
- international scale
  - options for the location of a new TNC factory/bank
  - UN disaster priorities post El Niño events

Refer to your specification for full details. Possible locations range from LDCs to MEDCs.

You are being tested on a number of assessment objectives, in particular your ability to apply your knowledge and understanding to an unfamiliar situation. If by chance the scenario is your own home area, you are not required to supply extra factual detail. It is the essential *general* geographical background that is important.

You will need to use your skills as a geographer to:
- analyse a range of data and geographical information
- select from and use evidence to make a synopsis of information in order to make decisions and to solve problems
- synthesise and evaluate data from the resources in order to explore their significance, impact and importance
- draw on your geographical knowledge to support your arguments

Achieving effective synoptic assessment

The specification
- Use the specification to find guidance on developing synoptic skills.
- Apply these principles to unfamiliar tasks in new situations.
- Make links between AS and A2 units.

Make the most of question papers
- The rubric on the papers reminds you of the need to show understanding of the connections between aspects of geography (e.g. between physical and human).
- Look for questions with synoptic triggers (e.g. demonstrating interrelationships).
- Look for high-level command words (e.g. synthesise, summarise and analyse).
- Be prepared to consider issues on a variey of scales.

Hitting the targets on the mark scheme
The mark scheme is organised in levels and includes credit for synopticity.
- Introduce background vocabulary to show your understanding of the geographical issues.
- Mention appropriate parallel examples you have studied, to inform your arguments.
- Look for overarching concepts, such as sustainability.

Convincing the markers
Synopticity has to be recognised by the markers.
- Show clearly how you have analysed, synthesised and interpreted the resources.
- Work through the exercise, but remain aware that synopticity is being rewarded and so incorporate relevant ideas from your course.

Decision making

The specification suggests you should look at:
- the process of decision making and problem solving
- the politics of decision making
- the psychology of decision making
- how economics affects decision making

- the environmental context for decision making (note the increasing emphasis on sustainable development) and the important role of environmental legislation
- the sociology of decision making (the movement from top-down to bottom-up community-orientated decision making)
- impact assessment and monitoring of decisions (potential and actual)

The specification seems very daunting but it helps you to think of these issues in real-life situations. There is no substitute for practical involvement in issues analysis to give you confidence. Such involvement might include:

- a visit to a planning enquiry about a local issue
- a debate between a number of players with differing views about a key issue
- a piece of fieldwork research about an issue
- a fieldwork monitoring exercise looking at the results of a decision

The next stage is to examine a range of question papers to see how they are designed to test your synopticity. In the Questions with Answer Guidelines section of this book, two new issues analysis exercises have been deconstructed to provide you with a writing framework to show your synoptic ability. Familiarise yourself with the demands of the mark schemes to see how much importance the examiners place on synoptic skills.

The synoptic assessment is worth 20% of your final grade, so you need to spend approximately 20% of your time working towards it.

Issues analysis strategies

(1) During your course — long term
- Become familiar with the specification framework.
- Practise decision-making techniques during the course — tabulation, matrices, scaling, weighting, ranks, environmental impact and assessment.
- Practise decision-making analysis — values and opinion analysis, option analysis and cost/benefit.
- Practise interpretation of decision-making resources — OS maps, data maps, photographs, tables, graphs, text and statistics focused on an issue.
- Practise decision-making execution. Try different scenarios.
- Get to grips with synopticity and what it means.

(2) Pre-release resources
- Ask your teachers to help you analyse difficult resources.
- Work on individual resources, especially OS maps:
  - What does each resource show?
  - How can it be used to provide information?
  - What tasks might be set?
  - What geographical skills might be required?
- Make an 'option fact file' — learn and synthesise all the details about each scheme, option or site.

- Do background reading on the geography of the issue but do *not* decide on possible questions.

(3) The exam paper

This comprises all resources, tasks or questions and a 'letter of instruction'.

- Read the letter very carefully as it is there to help you (e.g. with structuring your answer).
- Match resources to tasks. Plan for 5–10 minutes. Check that all the resources have been used.
- Identify what is needed for each task. Decide which techniques to use, if any.
- Work out what information is needed to answer each task.
- Quote detailed evidence from the resources.

Timing and tasks

- *Plan* for each task.
- When writing up:
  - match time to task and marks allocated
  - concentrate on the precise rubric and command words and do *exactly* what each question demands
  - refer back to the letter
  - quote *direct* evidence
  - make *brief* comments on parallel situations, especially from your fieldwork
- Use decision-making techniques *where appropriate*.
- Remember — structure, evidence and a comprehensive coverage of all questions.
- Always watch your quality of written communication.
- Allow 5–10 minutes for a final check.

Synoptic understanding

Synoptic understanding should be *flaunted* if you are to achieve the top-level band. Try to cross-link with similar examples you have studied, while remaining focused on the task set. Do not go into 'case study mode'.

Quality and cohesion

Quality and cohesion of the whole report is worth up to 10 marks. This includes:

- following a sequence of enquiry
- using evidence effectively
- writing in a structured, logical format
- providing a report fit for purpose, i.e. using bullet points, tables and techniques effectively and appropriately
- showing good use of geographical terminology
- demonstrating effective synopticity

Synoptic Guidance

This section is designed to help you develop issues analysis skills. Many of the techniques discussed are used during Unit 3 fieldwork and so some might be familiar to you.

When handling real-life issues, decision-makers have a range of tools that they can use to help reach a decision, or to provide a solution to a problem. Decisions vary in complexity and depth. One of the skills is to find the most appropriate technique. The skills and techniques in this section can be used to enhance your understanding of an issue and are extremely useful when handling the pre-release resources for the unit test.

This section is divided into the following topics:
- **issues analysis skills and techniques**
- **analysing resources**
- **evaluating options**
- **investigating values and opinions**
- **monitoring techniques: environmental impact assessment**
- **handling pre-release resources**

Experience with unit tests so far (including exercises on Bassett's Pole, Wharfedale and Breton Bay) has shown that students who build up their 'toolkit' of skills and techniques over their course are far more confident in an exam situation.

This section is designed to show you what is available and to give you practice with a variety of approaches. Rather than work through the whole of the section, try the exercises in context, for example using weighted tables (see p. 19) when revising coastal management (AS Unit 1).

Issues analysis skills and techniques

Decisions vary in complexity and depth. Therefore, there is a variety of tools or skills that can be used in order to reach a decision, opinion or conclusion. In an exam, you might be asked to:

- rank options in terms of their suitability against agreed criteria
- choose or justify the best or worst scheme
- decide on an order of priority for development
- evaluate the social, environmental or economic impact of a proposal
- assess advantages and disadvantages or costs against benefits
- decide where option X should be located with reference to surrounding sites

Tables 1 and 2 provide a brief summary and explanation of both basic and more sophisticated analytical tools that are available.

Table 1

Basic tools	Brief description
Tabulation	Simple summaries of information into themes or topics; tables can be a mixture of text and numbers
Scaling	Allows comparison between factors or variables; good for quantitative data that are difficult to judge in terms of the 'bigger picture'
Ranking	Provides a means of ordering options or schemes, so that priorities can be identified
Weighting	Gives additional significance to selected criteria, so that their impact or qualities are more pronounced in the final ranked score
Matrices	More complex summaries of data tabulated as grids; these show the relationships between individual variables

Table 2

More sophisticated tools	Brief description
Cost/benefit analysis (CBA)	Assesses the desirability of a project by examining the overall socioeconomic benefits to a community against the socio-economic costs of development
Sieve mapping	A spatial method for 'sifting' sites and options; projects or locations that are unsuitable are rejected at an early stage in the process
Critical path analysis	Often called 'decision trees', these may be represented as flow diagrams; they represent time-lines of activity

Tabulation

Tables are useful for summarising information about a complex issue into more manageable themes or topics. Tables could be used in the exam when you are asked to 'summarise' or 'compare', or to consider 'options' and 'opinions'. Make sure that:

- you use the correct framework for the question, with relevant headings
- you provide comprehensive coverage of the information required
- within the table, the language is sound and written in good geographical prose, using a maximum of two or three sentences for each point
- you remember that columns or rows can be added to tables to display information on scaling and ranking

Setting up tables

- Tables should be landscape (or over two pages) to give more space for writing.
- Select and identify the criteria to be used as headings — these must link exactly with the focus of the question or problem.
- Divide up the space so that it is equal for each option.
- Provide clear evidence, using supporting figures and geographical terminology.
- Make your tables neat, legible and logical.

Tip Tabulation:

- can lead to serious brevity!
- can limit the evaluation of data — it may be a good idea to provide a summary evaluation at the bottom of each column.
- is time consuming — is it really necessary?

Examples 1 and 2 below were produced by students when reviewing sites for four possible regeneration schemes in central Glasgow.

Example 1

Site	Location	Size and topography	Characteristics		Adjacent land use
			Good points	Bad points	
A	Directly south of plantation, approx. 0.75 km north of Pollokshields; 0.25 km south of the River Clyde; adjacent to the main ring road; 1.25 km away from Central Station and the CBD	0.6×0.25 km $= 0.15$ km^2 Flat	Good access from main roads; close to the CBD and riverside; largest site	Existing buildings on site may be expensive to clear; church on site; local opposition	Residential areas directly north of plantation; old works to the east and south; possible residential areas to the west; Pollokshields is residential

Site	Location	Size and topography	Characteristics		Adjacent land use
			Good points	Bad points	
B	In the centre of Govan, surrounded by residential land (e.g. Drumoyne to the west and south); approx. 0.3 km north of Bellahouston Park; 1.4 km south of R. Clyde; adjacent to main ring road; 4 km west of Central Station and CBD	0.3 ×0.35 km = 0.105 km^2 Slight slope to the north of the site	Good access from the main ring road; adjacent to the railway line	Existing buildings on site (demolition costs); not riverside location; fairly small site; not flat	Residential areas — Drumoyne to the west; Craigton to the south; old Victorian land to north and east; stadium 0.5 km to the east
C	Located on the riverside, above the Clyde tunnel (A739); 0.75 km north of Greater Govan; 1.3 km north of main ring road and railway line; 4.5 km northwest of Central Station and CBD	0.3 ×0.35 km = 0.105 km^2 Slight slope to the north of the site	Good access from both sides of the river as close to the Clyde tunnel; good access from ring road	Furthest away from CBD and railway; tunnel may lead to structural problems; sewage works 0.25 km to the west	Sewage works 0.25 km to the west; works 0.25 km to the east; residential areas to the south; Elder Park (green space) 0.2 km to the southeast
D	Located on the riverside with Prince's Dock adjacent to the west; 0.4 km north of the main ring road and 0.75 km north of the railway lines; 2 km west of Central Station and CBD; SECC located on the opposite river bank	0.4 × 0.3 km = 0.12 km^2 Flat	Clear site — no demolition costs; Prince's Dock is a potential marina; reasonable road access	No direct rail access; land could be expensive as it is already cleared	Plantation residential areas to the east; industrial land to the south; SECC located on river on the opposite bank; Prince's Dock directly to the west

- The criteria chosen as column headings match the information required by the site review.
- Each of the four sites is given a similar depth and extent of treatment.
- The description of each location is detailed and supported by figures.
- Elements of analysis are included in the tabulation (e.g. calculation of site area)

This tabulated response would gain marks in the top band.

Example 2

Environment

Site	Score of suitability and reason
A	Too far from the river; surrounding area not suitable — dirty, busy M8 and factories
B	Would improve the built-up area, but not close to the river and busy M8 not suitable
C	Very close to the river; one cross-river link already; large vacant area to use, but no large roads nearby; already near to the similar Elder Park
D	Large area; Prince's Dock and river links could be used and exploited; possibly a little too far to the east?

Housing and light industry

Site	Score of suitability and reason
A	Already industrial area which could be regenerated; few houses nearby; would people want houses so close to M8?
B	Lots of poor housing nearby; good transport links could be improved further; if jobs were provided, they would be close to possible employees; good access
C	Bad access; would obscure links across river
D	Bad access; few houses nearby

This example is much weaker:

- It lacks descriptive detail and precision and does not use a suitable range of criteria for site evaluation.
- Insufficient criteria are selected. Environment, housing and industry schemes will have very different requirements.
- The descriptions lack detail and specifics, for instance locations and distances.
- Tables should not raise questions, but should be used to state facts.
- It seems to run out of steam towards the end — the explanations for sites C and D lack details and justification.

This tabulated response gained marks in the lower band criteria.

Tip Summary tables are a useful way of sifting through the evidence from the pre-release resources. Well-developed tables provide a synthesis or synopsis of the key points about each option or site. Don't repeat the information in your table in a written explanation alongside. This common mistake in exams simply wastes time and energy.

Example 3

This example is taken from a summary table about quality of life in Jakarta, Indonesia.

Location	Climate	Infant mortality	People per doctor; life expectancy	Telephone lines per 1000	CO_2 emissions	GNP per capita	Total
Jakarta	26°; 80%	65	7143; 53	1.3	170 468	750	23

- The information supplied is far too brief.
- Units are missing.
- The meaning of the final column is unclear.

At first sight, this response is an almost meaningless grid of numbers with no explanation. However, some of the criteria chosen are useful (i.e. those that provide socioeconomic and environmental data). The evidence base would be improved with some explanation.

Scaling

Scaling permits comparison and analysis, especially of quantitative data. It allows some sort of standard structure or relative positioning of the data to be incorporated into the decision-making process. There is nothing magical about scaling, although it often involves some sort of 'fiddle factor'!

Numerical data are the easiest to scale, although scaling can also be applied to more qualitative data. Most scaling methods involve personal, subjective judgments or measurements that may introduce some bias into the analysis.

Nominal scaling

This is the simplest level of scaling. Data take the form of categories, for example male or female, arable or pasture. Observations or data are allocated to particular categories. These data cannot be ordered.

In Table 3, three options (A, B and C) are given for a road improvement scheme. The results from a consultation document reveal how each option faired in terms of three simple objectives:

Table 3

Option	Improvement in traffic flow	Aesthetic improvement	Increase in safety	'Yes' totals
A	Y	N	Y	2
B	N	N	Y	1
C	Y	Y	Y	3
Y = Yes, objective met; N = No, objective not met				

Option C is the preferred choice for the new road scheme.

Tip Be prepared to be flexible using this technique, and don't be afraid to adapt the process to help you reach decisions. Remember there are many types of nominal data that could be scaled, such as religious background, gender and ethnicity.

Ordinal scaling and ranking

Most tabulated data presented for use in decision making or issues analysis is ordinal. Such data can be put in order or ranked in some way, for example income levels, A-level pass rates and the number of people in a family. This approach is also useful when absolute values are not known, but relative positions on a scale can be used.

Using the same example as above, the 'yes'/'no' responses have been converted to an ordinal scale (Table 4):

Table 4

Option	Traffic flow	Aesthetic improvement	Increase in safety	Rank total
A	3	0	2	5
B	0	0	2	2
C	3	1	2	6
Ranking: 1 = low priority, 2 = medium priority, 3 = high priority				

Option C is still preferred, but Option A comes a close second.

Ranking of data is especially useful when dealing with resources such as large tables or a great deal of statistical information. Ranking should reveal patterns in the data, making comparisons easier, and permitting a choice to be made between options. Ranking is usually undertaken in terms of the suitability of the option against a set of agreed criteria or it might be used to look at the best and worst locations. Table 5 shows a way of setting up a basic ranking table:

Table 5

Option	Criterion 1	Rank	Criterion 2	Rank	Criterion 3	Rank	Total rank	Final position
A		1		2		4	7	2
B								

- The boxes below criteria 1–3 can include short written statements, backed up by supporting data. These boxes can contain both qualitative and quantitative data.
- You need to explain what is high and what is low ranking — most or least damaging, best or worst site and so on.
- The total rank is the sum of the rank scores for each option. An average rank could be used.
- The final position score allows the reader to make simple comparisons.

Example

Tables 6 and 7 show basic demographic and social data for four Asian countries and their capital cities.

Table 6

	Malaysia	Singapore	Indonesia	Thailand
Population, 2003 (millions)	23	4.0	235	64
Average annual population increase (%)	1.9	0.8	1.5	0.8
Infant mortality (per thousand live births, 1980–2000 average)	8	2	46	25
Population per doctor, 1994	2564	725	7143	4762
Average life expectancy at birth, 1990s	65	72	53	63
Population living in urban area in 1998 (%)	57	100	39	38
Number of cars, 1991(thousands)	2000	285	1416	825
Education spending as % of GDP, 1994	5.3	3.3	1.3	3.8
% secondary school enrolment, 1997	65	72	52	31
Adult (>15 years old) literacy, 2001 (%)	88	92.5	88	97

Table 7

	Kuala Lumpur	Singapore	Jakarta	Bangkok
Index of living costs, 1996 (London = 100)	88	109	89	88
Population (millions), 2000	1.1	3	8.2	5.9
Vacancy rate for prime office space (%)	9	7	12	24
Occupancy costs for prime office space ($US per m²)	650	1400	250	225
World quality of living index, 1996	53	9	81	85
Carbon dioxide emissions by industry, 1991 (tonnes)	61 169	41 293	170 468	100 896

Note that a high number in the world quality of living index means a low quality of life.

The task was to make an assessment of quality of life for people living in the capital cities (likely to be slightly different) of the four countries shown.

Tip Remember:
- not to be afraid to discard irrelevant or out-of-date data (e.g. population)
- to make sense of the data (e.g. 2 million cars in Malaysia, not 2000!)
- that this type of question implies that you should use ranking
- to divide the data into environmental and socioeconomic quality of life indicators

Student response

Capital city	Climate	Employment	Quality of life	Carbon dioxide emissions	Total rank	Final position
Jakarta	3	2	3	4	12	4
Kuala Lumpur	2	3	2	2	9	1=
Singapore	4	1	1	3	9	1=
Bangkok	1	4	4	1	10	3

This is a weak bottom-band response.
- A limited range of criteria has been used. 'Climate' and 'Employment' are not explained.
- The criteria are poorly selected — quality of life should be the final focus, and include 'Environmental' and 'Economic'.
- No evidence or explanation is provided for the ranking figures.
- The original table did not include the final two columns, which provide a good summary.
- The selection mixes both country and city data.

Tip
- When ranking, it is *critical* to select the correct criteria, preferably with a well-justified range. They must all be fit for the purpose.
- *Always* include *interpreted* evidence from the data or resources supplied — don't just lift information directly.
- *Always* explain *exactly* how the scoring system works, and show how the ranks are derived.
- Remember that quantitative data are much easier to rank than qualitative statements.

Thinking about the tips above, use the data in the tables on page 17 to develop your own fully evidenced ranking table.

Weighting

Weighting is a natural progression from the previous forms of scaling. It gives additional significance to selected criteria, so that they impact more in the final ranked score. Weighting is normally based on mathematical 'fudges', such as doubling or trebling one or more scores in the table.

In the following example, a student was presented with a range of coastal protection schemes.

The project brief identified a need to provide a cost-effective and long-term solution for a 1 km stretch of busy amenity beach on a high-energy coastline, which is not low-lying. The question implied that weighted scores should be used to reach a decision.

Tables 8 and 9 are copies of those developed by the student.

Table 8

Coastal scheme	Description	Approximate cost (2000)	Typical lifespan/years	Possible outcomes
Concrete sea walls	Popular at sea-side resorts (promenades)	£5000 per metre	50–75	Can be undermined, by waves (toe scour) but good in amenity sites if linked to other protection
Revetments	Lower cost defences, in less developed coastlines	£2000 per metre	25–30	Effective at absorbing wave energy, but do not give full protection
Groynes	Popular on tourist beaches, every 100–200 m	£10 000 each (wooden)	25–40	Trap sediment on tourist beaches, but result in sediment starvation down-drift
Gabions	Boulder-filled wire cages, often behind revetments	£500 per metre	10–30	Rapidly break up in high-energy environments
Earth embankments	Flood defences along low-lying, low-intensity coasts	£1000 per metre	Less than 10 years before repair	Protect farmland in low-lying areas, but easily breached by storms
Rip-rap	Piles of huge boulders, often patching up failed schemes	£500 per boulder	Short term	Huge mass protects in high-energy environments
Beach nourishment	Gravel and sand dumped by local authorities on tourist beaches	Cheap, but labour costs high	Short term	Easily moved by offshore currents and longshore drift, so needs regular feeding

Table 9

Coastal scheme	Cost ranking (weighted)	Lifespan ranking (weighted)	Outcome ranking	Total score (final rank)
Concrete sea walls	7 (21)	1 (2)	2	25 (4)
Revetments	6 (18)	3 (6)	4	28 (6)
Groynes	3 (9)	2 (4)	3	16 (1)
Gabions	4 (12)	4 (8)	6	26 (5)
Earth embankments	5 (15)	5 (10)	7	32 (7)
Rip-rap	2 (6)	6.5 (13)	1	20 (2)
Beach feeding	1 (3)	6.5 (13)	5	21 (3)

- Ranks run from (1) highest (cheapest/most effective/least impact) to (5) lowest (most expensive/least effective/most impact).
- The description is not ranked because the information is qualitative and subject to bias.
- Costs were calculated for a 1 km stretch and then ranked. These were considered the most important factor and were weighted (×3).
- The lifespan was weighted (×2), after the initial ranking.
- Possible outcomes are difficult to rank, because they are qualitative. The rank is based on a 'best guess'. No weighting has been applied.
- The final summary column is the most important.
- Using the ranked table, it is possible to conclude that groynes are the most effective option for this location and earth embankments are the least effective. Do you agree or would you come to a different conclusion?

Tip Weighting can introduce bias, as the justification for individual weightings may be based on patchy knowledge or incomplete evidence. Weightings that are applied must be both explained and meaningful. Don't feel you have to rate, rank or weight every element of the data. There is a temptation to do this to try to look more 'scientific' or achieve better marks. Resist this, as it can lead to spurious conclusions.

Bipolar analysis

You may be familiar with bipolar scaling from questionnaires. Usually, two extremes of opinion, attitude or variable are represented along a continuous line, which is subdivided into categories. This technique is especially useful if the data are qualitative or subjective.

You have to:
- choose a range of suitable **descriptors** for the task
- assign an appropriate number to each descriptor

Bipolar scales may operate from as much as –10 to +10, but more usually from –3 to +3. In some cases, an ordinal 0–5 scale may be more appropriate.

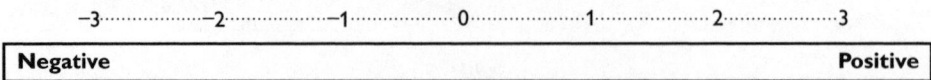

–3 ·············· –2 ·············· –1 ·············· 0 ·············· 1 ·············· 2 ·············· 3

Negative	**Positive**

You can choose the extremes of the scale and the intervals to suit your purpose.

The bipolar technique can be used to describe your own attitude or feelings about a particular scheme, project or place. It can also be used to show how a group of people or users feel, or how other interested individuals respond.

Have a look at the examples below:

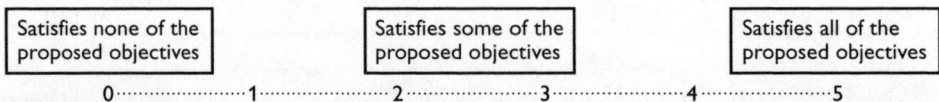

Satisfies none of the proposed objectives	Satisfies some of the proposed objectives	Satisfies all of the proposed objectives

0 ·············· 1 ·············· 2 ·············· 3 ·············· 4 ·············· 5

The above case illustrates some key problems:
- The descriptors do not describe all of the intervals (or scores) used on the scale.
- The descriptors are rather vague, making interpretation difficult.

It could be reworked into the following:

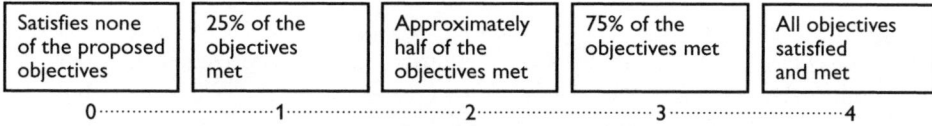

Satisfies none of the proposed objectives	25% of the objectives met	Approximately half of the objectives met	75% of the objectives met	All objectives satisfied and met

```
0 ························1························2···········3··························4
```

If data from different groups or projects are difficult to interpret, a 'profile chart' may be useful. The annotated chart below (Figure 1) is for a small area before and after regeneration.

The chart works well because there are only two lines. However, while it is visually attractive, it is more time-consuming to construct than a table.

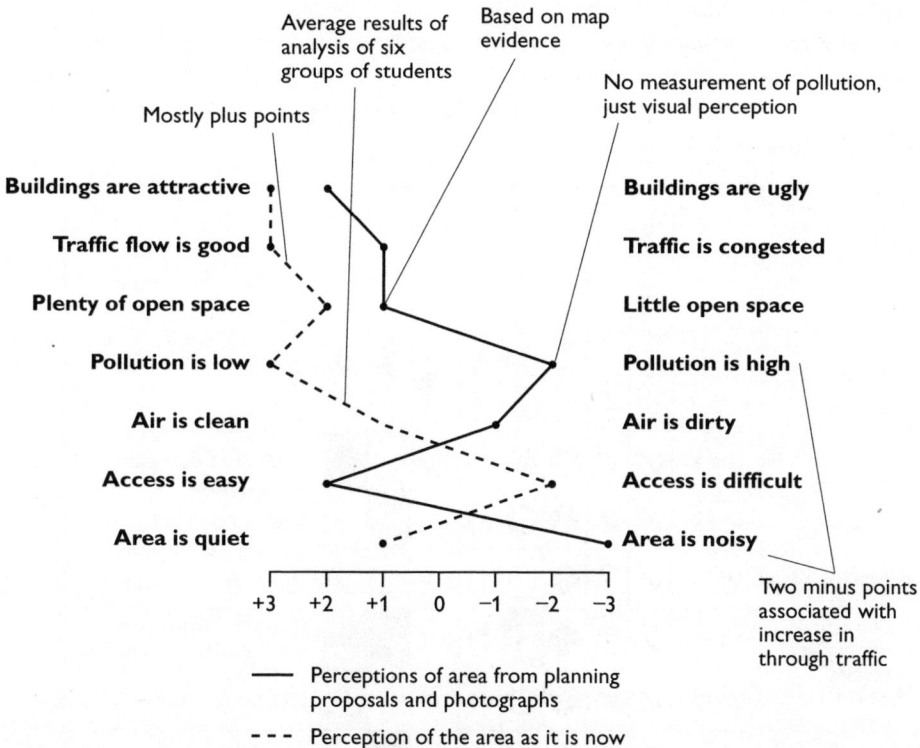

Figure 1

Tip The bipolar approach is non-scientific and open to errors. Small group discussion in class may allow 'calibration' of the scale so that everyone understands the descriptors being used. The use of bipolar scales improves with practice, avoiding decisions that are crude, simplistic and unconvincing. Remember that the results obtained are tentative indicators, not definite quantities.

Matrices

There are many different types of matrix. However, they should all be designed with the clear purpose of providing a way of summarising, a checklist, or a tool for assessing conflicts.

A matrix is usually a diagram or grid constructed to cross-reference the effects of options or conflicts against another set of criteria. The rows represent one set of data; the columns the other set. The intersecting cells represent the relationship between individual variables in each set, at that point.

Tip Matrices are very adaptable techniques that can be of use in all stages of issues analysis and decision making. They are especially useful in bringing order to lots of data — to rank or rate the data, compare alternatives, summarise findings and compare pay-offs. The matrix really comes into its own in assessing the extent to which the impacts of policies or projects intended to do one thing affect others that are rarely considered.

Figure 2 is a basic matrix in which:
- the column headings represent activities and users
- the row headers show the various alternatives of the project/plan/programme
- the cells of the table contain figures, descriptive text, symbols and colours — a range is shown for illustrative purposes

	Pedestrians and joggers	Sun-bathers	Dog-walkers	Rollerbladers and cyclists	Children	Anglers	Teenagers	Environment	Community	
Site 1	✓	✓	⑧	●	0.1	3.24				No impact detected
Site 2	✗	✓	❶	○	0.3	4.57				Impact significant
Site 3	✓	?	❹	○	0.1	6.81				Will meet all objectives
Site 4	✓	?	❺	◎	0.1	0.32				Doesn't satisfy any of the proposed objectives
Site 5	✓	✗	❶	○	0.2	1.21				25% of objectives met
Site 6	?	✗	❶	◉	0.2	1.56				> 75% of objectives met

The highlighted cell indicates the interaction of 'rollerbladers and cyclists' with site 4

Figure 2

Tip It's often a good idea to set up the matrix with the paper turned landscape so that there is more space for detail. Remember to include a key to explain symbols or coding.

Checklist matrix

The checklist matrix is the commonest type. The relationship is plotted on two axes. Often no attempt is made to quantify cell entries, although there should be some indication of how decisions on cell entry have been made.

Cross tabulation matrix

A cross tabulation matrix is usually a simple mathematical matrix, enabling numbers to be added or multiplied together to give a final result, usually in the right-hand column. This type of 'checking' matrix can be easily put together as a spreadsheet.

Table 10 is part of a matrix produced by a student in response to an issues analysis question — an evaluation of proposed routes for a new road. Adding together the scores for each proposal gives a final result.

Table 10

Criterion	Proposal				
	A	B	C	D	E
Cost	8	−8	5	6	1
Impact on residents	2	2	0	−1	−3
Traffic flow	8	7	6	6	4
Pedestrian safety	−4	−6	−7	−3	−4
Impact on tourism	5	0	−3	6	0
Total	32	10	4	32	13

Key: 10 = most beneficial; 0 = no effect; −10 = least beneficial

This extract shows an impressive and comprehensive use of numerical scoring, but there is no indication of how the scoring links with the criteria. It would be improved by better evidence and a more comprehensive key.

Example 1: Florida case study

This case study looks at environmental restoration options.

- Sarasota Bay is located on the central west coast of Florida between Tampa and Venice (see Figure 3). The mangrove environment is bordered by a chain of coastal barrier islands.
- The coastal wetlands and seagrass meadows in the Sarasota Bay region have been significantly affected by dredge and fill placement from the Gulf Intracoastal Waterway during the late 1950s to early 1960s. Dredge material was placed in mangroves and shallow water bay bottoms, creating upland areas which were invaded by exotic vegetation.

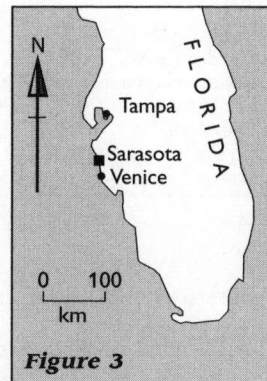

Figure 3

- The West Coast Inland Navigation District commissioned a study to examine environmental restoration projects for the coastal lagoon system to help determine the most feasible and environmentally productive alternatives for Sarasota Bay. Six islands will be evaluated as a part of the feasibility study, including Big Edwards Island (see Figure 4).

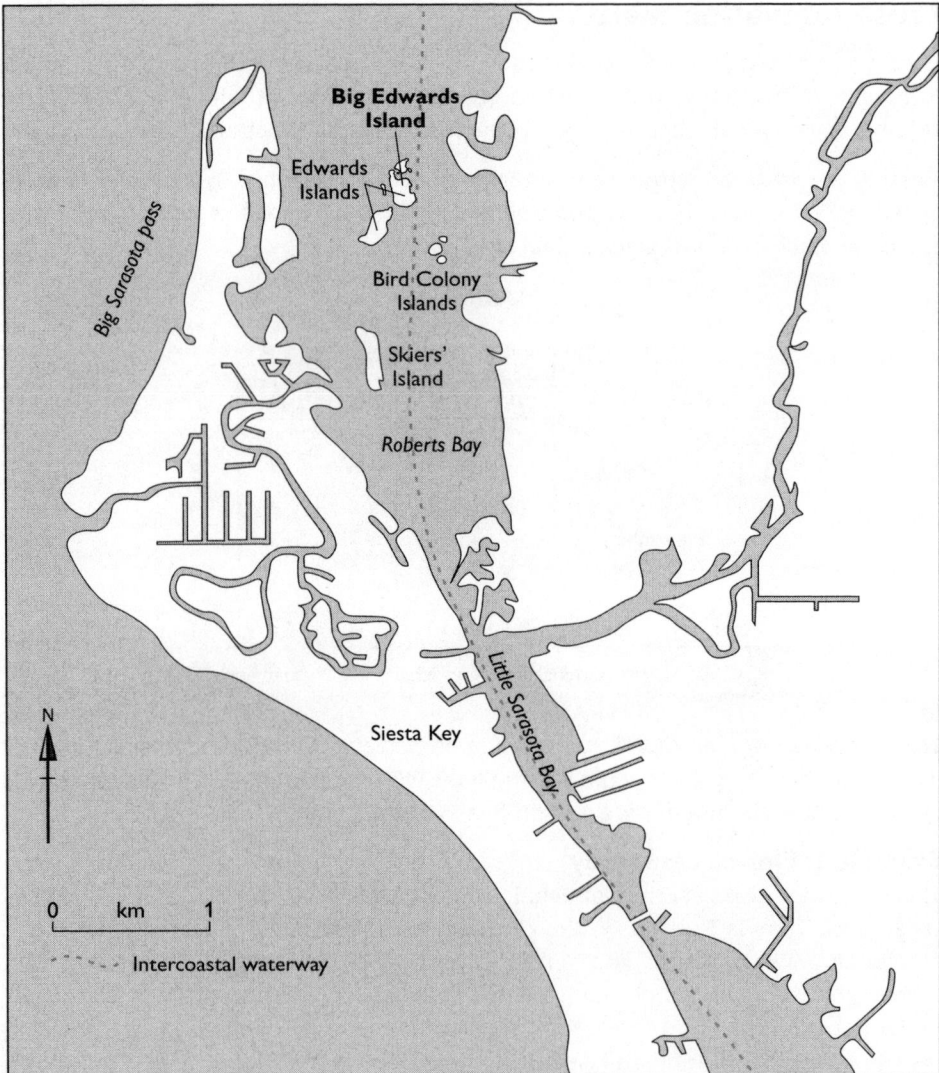

Figure 4

Figure 5 shows four concepts for Big Edwards Island. The options are evaluated in Table 11 (page 26).

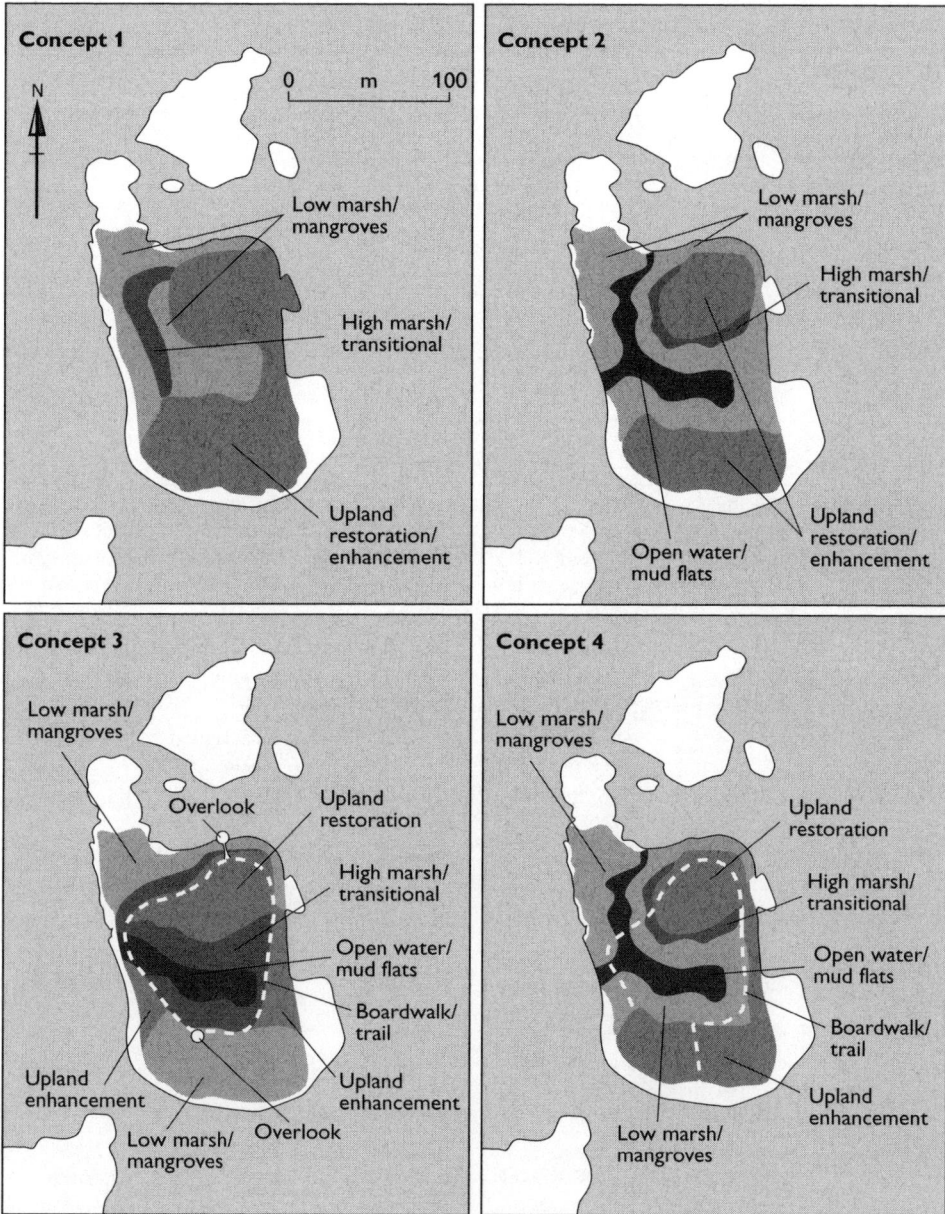

Figure 5

Use Table 11 (a checklist matrix) to develop a cross tabulation matrix to evaluate the *best* concept for Big Edwards Island on environmental grounds. You should redraw the matrix and insert scaling.

Table 11

Environmental factor	Concept 1	Concept 2	Concept 3	Concept 4	No action
Total area of habitat types created (acres)	4.4	4.3	4.1	4.0	0
Upland restoration (acres)	2.7	1.7	1.3	1.7	0
High marsh (acres)	0.3	0.2	1.0	0.2	0
Low marsh and mangrove (acres)	1.4	1.9	1.8	1.6	0
Tidal lagoon/ mud flats (acres)	0	0.5	0	0.5	0
Fish and wildlife resources	Creates potential nesting and migratory bird habitat Low marsh — potential fisheries habitat	Creates potential nesting and migratory bird habitat Low marsh — potential fisheries habitat Tidal lagoon creates feeding grounds for invertebrates, fish and shorebirds	Creates potential nesting and migratory bird habitat Low marsh — potential fisheries habitat	Creates potential nesting and migratory bird habitat Low marsh — potential fisheries habitat Tidal lagoon creates feeding grounds for invertebrates, fish and shorebirds	Continued degradation of uplands and low marsh by exotic vegetation
Removal of exotic vegetation	Yes	Yes	Yes	Yes	No
Shoreline erosion	No impact	No impact	No impact	No impact	No impact
Water quality	Improves — creates low/ high marsh wetlands	Improves — creates low/ high marsh wetlands	Improves — creates low/ high marsh wetlands	Improves — creates low/ high marsh wetlands	No impact
Recreation	Upland areas provide public access to the island for passive recreation	Upland areas provide public access to the island for passive recreation	Provides boardwalk, overlooks, and educational signage for public use	Provides boardwalk, overlooks, and educational signage for public use	Upland areas provide public access to the island for passive recreation
Navigation	No impact	No impact	No impact	No impact	No impact
Public acceptance	Moderate	Moderate	High	High	Moderate
Cost estimate ($m)	0.8–1.35	0.7–1.15	0.7–1.15	0.65–1.1	n.a.

Summary matrix

A summary matrix is a useful housekeeping device, for storing data and providing summary information. Often summary matrices have text rather than numbers, to make them more meaningful. Table 12 shows a summary of the views of different interest groups on a new town plan which had four options.

Table 12

Interest groups	Views	Main objections	Main ideas	Favoured option	% voting for
History Society	Conservationists — keen on history of the area	Redevelopment of historic town centre — loss of history	Old buildings are important to attract visitors	One	40
Chamber of Trade	Want to increase visitors and visitor access	Loss of passing trade if pedestrianised	Old buildings are uneconomic and dangerous	Two	38

Conflict (compatibility) matrix

In this type of matrix, usually only one project or programme is examined. The axes can be identical and ticks and crosses are often used. Figure 6 is from a conflict matrix of lake-users on the imposition of a 10 mph speed limit on Lake Windermere.

Figure 6

	Power-boat users	Swimmers	Windsurfers	Tour-boat operators	Lakeside walkers
Power-boat users		✗	✗	✓	?
Swimmers	✗		?	✗	✓
Windsurfers	✗	✓		?	✓
Tour-boat operators	✓	?	?		✗
Lakeside walkers	?	✓	✓	✗	

Activity:
✓ = compatible mix
✗ = conflicting mix
? = uncertain

Cost/benefit analysis (CBA)

This technique should really be called 'benefit/cost analysis' because that is what it does — assesses the desirability of a project by examining the overall social and economic benefits to a community against the social and economic costs of development.

An important aspect of CBA is that it attempts to use monetary values to establish the 'costs' and 'benefits' of a particular proposal, such as coastal or flooding management.

Example 1

The costs of implementing various hard flood defensive schemes are compared with the benefits of saving properties located on the floodplain for a number of different flood frequency events (Table 13). (Hard refers to the construction of walls or similar to contain the river under bankfull conditions.)

- **Cost** — the estimated price of flood scheme development
- **Benefit** — the damage prevented (insurance costs)

Table 13

Flood frequency	Cost/£m	Properties and businesses protected	Benefit (£m)	Benefit/cost ratio
1 in 10 years	2.4	58	1.3	0.54
1 in 20 years	2.5	105	2.1	0.84
1 in 50 years	3.0	210	4.2	1.40
1 in 100 years	3.8	375	5.9	1.55
1 in 150 years	4.6	383	6.0	1.30
1 in 200 years	6.9	392	6.1	0.88

If these data are plotted as a graph (Figure 7), it becomes easier to see which flood defensive scheme is the most cost-effective:

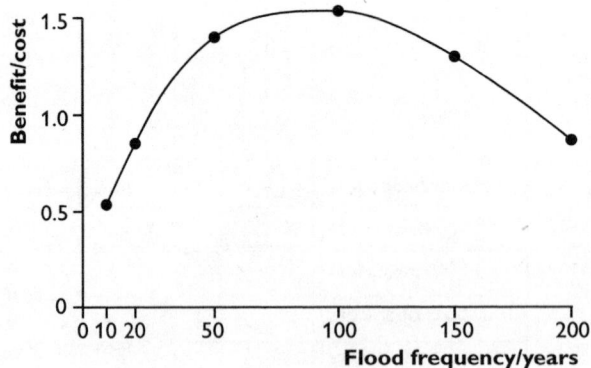

Figure 7

The point at which the curve is highest equals the greatest benefit derived from the least cost. Therefore, the most cost-effective protection scheme is at 1 in 100 years.

$$\frac{\text{benefit}}{\text{cost}} = \frac{5.9}{3.8} = 1.55$$

In reality, CBA can be more complex than the above example because some factors are 'intangible', that is, they cannot be readily priced or valued.

Conducting a CBA

Use the following steps:
- Draw up a matrix with three columns.

Column 1	Column 2	Column 3
List of elements that might be changed or individuals/groups that might be affected	Expected costs (negative elements)	Expected benefits (positive aspects)

- In columns 2 and 3 there must be a value for each cost or benefit. It is easiest to do this by using monetary values, but it is possible to use your own ranking or rating system.
- Add up the totals for columns 2 and 3.
- Work out the **utility:**
  total benefits – total costs
  A positive result shows greater benefits than costs.
- Calculate the cost:benefit ratio = benefits÷costs
  The higher the figure, the greater is the benefit.

Impact analysis table

If you are unable to calculate the actual costs and benefits for a particular project or proposal, then consider making a simpler, impact analysis table. In this, the columns can be headed '+' (advantages) and '–' (disadvantages) and details of each impact can be text.

Example

Table 14 summarises the costs and benefits of implementing hard flood defences in an English town.

Table 14

Economy		Environment		People and community	
+	–	+	–	+	–
Regeneration of house prices in previously 'unsaleable' locations	Implementation is expensive (depending on the option chosen)	Well-designed hard defences can enhance the aesthetic quality of the environment	Noise and dust during the construction phase	New jobs may be created during the construction and consultation phases	Local residents may be against potentially unsightly walls surrounding the river
Opportunity for new businesses to move into flood-protected areas	Public may think money would be better spent on local infrastructure and services	Construction of walls may create new habitats and ecological niches	Alteration of delicate ecosystem (including during the construction phase)	New, more attractive, riverside walks may be developed	Channelising can have adverse effects on downstream settlements in times of flood

Tip Remember that:
- all benefits and costs of a project should be measured in terms of their monetary value — this is not always easy, because some factors, such as health, don't have an easily calculated monetary value
- benefits and costs will change over time — this is often difficult to factor into the calculations
- there is a tendency to undervalue benefits and overestimate costs during CBA because costs are usually easier to quantify

Sieve mapping and analysis

Planners often use sieve mapping to identify areas where projects or developments can take place with least controversy and local objection. This simple, yet powerful, method requires the identification of constraints that would be incompatible with the proposal. These are then graphically represented, either on a map or GIS image, and 'overlaid' onto the site options. This idea of 'sieving' allows unsuitable areas or options to be weeded out early in the decision-making process. Eventually all layers are sifted to reveal an area or location which fits all the desired criteria.

The sieve overlay approach usually makes use of three types of information:
- site-based data — soils, vegetation, geology, drainage, topography, altitude
- administrative data — zoning, proposed boundaries
- location data — specific information about the proximity to the infrastructure (e.g. sewerage systems, gas and electricity services, settlement and roads)

Method
- Start by constructing a paper base map of the proposed site or development. The scale and area will depend on the option parameters. The base map should show the area of all options or sites.
- Next, draw various maps on transparent plastic (or tracing paper) — these are the different 'layers'. Each map should show a single criterion that is considered to be important in the site location. When the transparent layers are overlaid, option sites and areas can be filtered out.

Example
Eight new sites have been proposed for the relocation of Shrewsbury Town Football Club. The current Gay Meadow site is a town centre location, which provides easy rail access. However, there are issues with car parking and road access. The site is adjacent to the river and is prone to regular flooding, which also affects fixtures.

The potential relocation sites have been numbered and placed onto a base map, which includes a scale and north arrow. The following sieve layers have been constructed:
- Coarse sieve (Figure 8) — this shows areas of land that are low-lying and prone to flooding, based on a 1 in 50 year event.

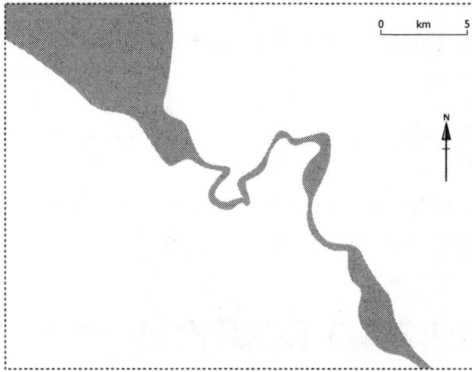

Figure 8

- Middle sieve (Figure 9) — this shows the main transport and access roads into the town. They are coded according to their size and therefore the volume of traffic they can handle.

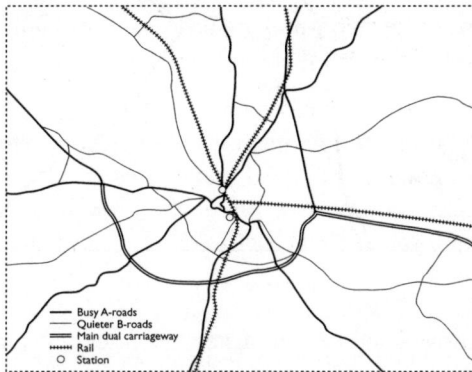

Figure 9

- Fine sieve (Figure 10) — this shows areas of residential and commercial land-use.

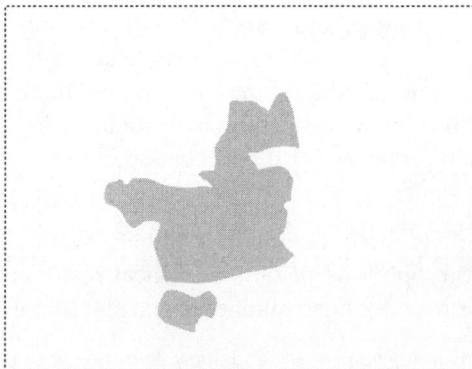

Figure 10

The combined result is shown in Figure 11.

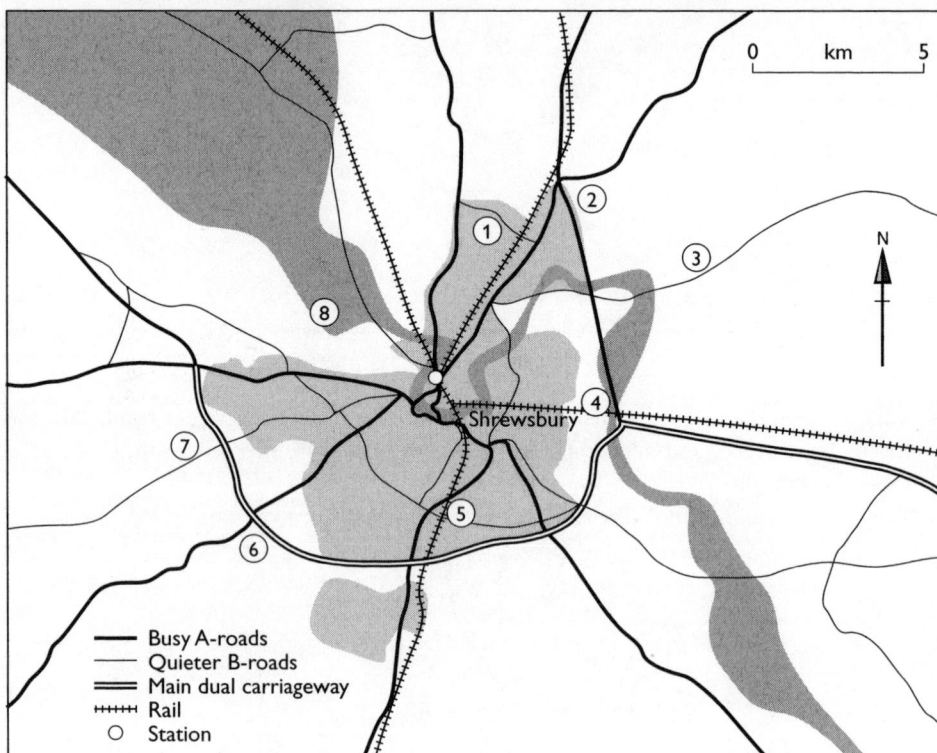

Figure 11

- The coarse sieve removes sites 8 and 4 because they are located on the flood-plain.
- The middle sieve indicates that site 3 is inaccessible and can only be accessed from a minor road. Sites 1 and 2 are some distance from the dual carriageway, so it would be costly to introduce new access roads.
- The fine sieve shows that sites 1 and 5 are within the built-up areas of Shrewsbury — noise and traffic congestion would be obvious issues. Site 2 is on the periphery of the current built-up area. In the future, this land might be developed for more residential capacity, thus impacting on the new stadium.

Based on the sieve analysis, sites 6 and 7 are possible options. Detailed ground surveys, together with full site assessments, would be the next stage in the issues analysis process. Brownfield, greenfield and greenbelt issues need to be explored. Questionnaires about the opinions of fans and local residents could be used as supplementary evidence.

Tip Overlays are time-consuming to prepare and apply. Working with many different layers is complex. Using only simple sieves, consideration of all the factors important in a new project is almost impossible.

Critical path analysis

Critical path analysis (CPA) is a management tool that was developed to identify the best route (critical path) in a chain of sequential and complex decisions and work out how long it would take to complete this critical path.

Think of critical paths as flow diagrams or decision trees. Often they represent time-lines of activity.

Example

Four 'green' or alternative energy generation schemes have been proposed for a hypothetical area of countryside. Four possible sites (A–D) have also been identified (see Figure 12 and Table 15).

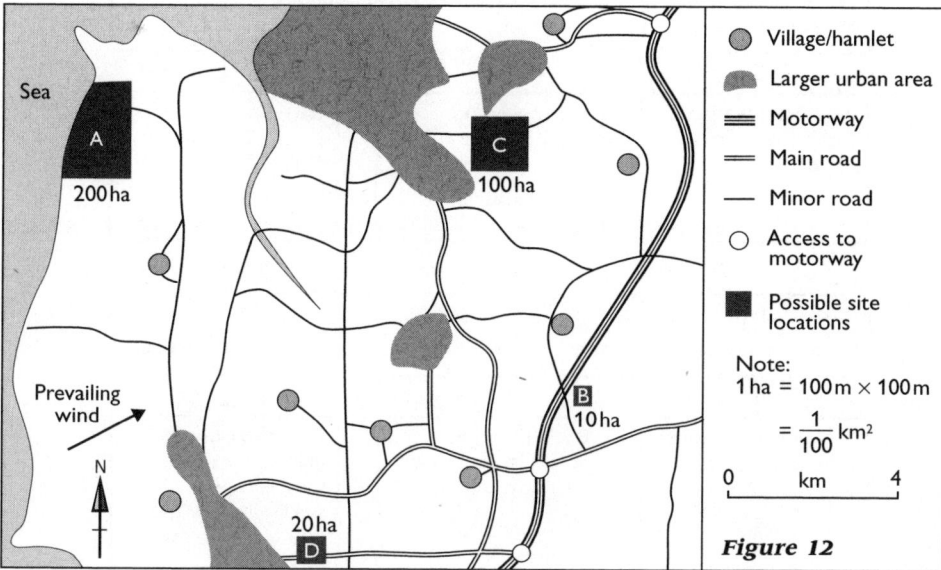

Figure 12

Table 15

Site	Location	Area (ha)	Altitude (m)
A	Coastal	200	70
B	Adjacent to motorway	10	30
C	Inland, close to a small town	100	120
D	Close to motorway and built-up area	20	90

Basic CPA is performed to identify:
- which sites are most suitable
- the most appropriate renewable energy scheme

Wind turbine farm

Needs an exposed location, annual wind speeds averaging 5.5m sec^{-1} and a large area

A — coastal site, getting prevailing southwest wind; rural, plenty of space; potential loss of amenity ✓✓	B and C — too inland and would not be windy enough ✗✗	D — quite coastal, but close to a built-up area; problems with space and noise ✗

Photovoltaic conversion plant

Converts solar radiation into electricity; an out-in-the-open site with few obstacles is ideal

A — space available but may be a coastal eyesore ✓	B probably too small ✗✗ D may just do ?	C — close to the town, but provides direct market for electricity ✓

Biofuel plant

Crops are used as fuel; transport of raw materials from the local area is crucial

A — unpopular with visitors to coast ✗	B and D — good access to motorway for raw materials ✓	C — too close to built-up area (problems with smells); high altitude means visible scar on landscape ✗✗

Waste-to-energy incineration plant

Incineration of domestic waste produces electricity; a plentiful supply of rubbish is necessary

A — coastal eyesore; inaccessible for transport of wastes ✗	B — small site, lacks space ✗ D — small site and relatively inaccessible for waste transport ✗	C — inland, but downwind of town; good transport links ✓

Possible solutions are shown in Table 16. Only one plant can go to each location, so the solution using CPA is circled. A and C are automatic choices, leaving D as the biofuel plant as it is too small for a photovoltaic conversion plant, which therefore has to go to B.

Table 16

Scheme	Possible solution (1)	Possible solution (2)
Wind turbine farm	(A)	–
Photovoltaic conversion plant	A	C or (B)
Biofuel plant	B	(D)
Waste-to-energy incineration plant	(C)	–

Tip CPA can be difficult to perform, especially for complex projects. Know your limitations and don't get bogged down with intricacies. It may be better to go for the simplified flow-diagram approach.

Summary of techniques

You should practise the techniques covered in this section in different scenarios to see how they support your analysis. They are all time-consuming and have to be very well done to be of use in evaluation. You will find them very useful when working on the pre-release materials, helping to make sense of the resources, options and opinions. The usefulness of the techniques is summarised in Table 17.

Table 17

Technique	Advantages	Disadvantages
Tabulation	• Good for summarising information quickly • Simple • Good for sorting pre-release materials	• Provides a chance to duplicate information in the table and in the text • Can lead to lists of meaningless numbers, with no explanation or information in too brief a format
Scaling	• Used in many aspects of decision making and may act as an important step towards further analysis • Might be the only option for sorting out preferences between people	• Some items are qualitative and difficult to scale in numerical terms • Relies on 'best guesses' and should not be used statistically • Often a personal interpretation of the facts available
Ranking and weighting	• Powerful tools in understanding the relative importance of aspects of a project • The result is often more convincing if an element of ranking or weighting has been applied successfully	• Weighting invariably introduces a degree of bias as 'weights' are based on incomplete knowledge and evidence • Ranking is often based on feelings and not necessarily on scientific data

Technique	Advantages	Disadvantages
Matrices	• One of the most useful and adaptable techniques available to the decision maker • Help bring order to many elements of data and reveal relationships that might not otherwise be noticed; these might also then be tested statistically • Generally simple to construct and look neat and efficient	• Can be complex, both to construct and for the end-user to understand • There is a wide range of matrices available, some of which have dubious utility (e.g. 3D and concentric circle matrices)
Cost/benefit analysis	• Industry-standard technique, which carries credibility • Very simple technique to grasp in terms of theory • Many examples, especially from large projects	• Normally relies on the identification of monetary values in terms of 'costs' and 'benefits'; these are often complex and not easy to obtain • Tends to lead to bias because costs are more readily expressed in monetary terms, compared with 'intangible' benefits
Sieve mapping	• A logical and easy to understand visual technique • Especially good for decisions that involve multiple-site options in different geographical areas • Can be linked to GIS analysis	• Maps may take a long time to research and produce • Too many maps make the process complex — most projects require the production of many 'layers'
Critical path analysis	• If time is limited, simple flow charts can help summarise a range of options for different projects • Can help you break down the decision into manageable chunks so you can think about it more logically	• A full CPA is complex and time-consuming; often not relevant or necessary in geographical decision making • Commercially, CPA is done through computer modelling

Analysing resources

You should expect as part of any decision-making or issues analysis exercise to be confronted with a wide range of resources. These include:

- **written resources** — books, articles, data sets and reports from an official body
- **maps** — OS, land-use, historical, geology, street plans
- **photographs** — include digital images, satellite images and aerial photos
- **diagrams** — another method for displaying various types of raw and processed data
- **questionnaire data** — raw data in the form of questionnaire returns or tabulated, processed information

- **graphs** — many different ways to represent data
- **interviews** — qualitative interview records, from which you have to make decisions or judgements

Tip Analysis of resources is a fundamental skill in issues analysis. In a court of law, it is essential that any decisions made concerning the accused are accurate, so that a fair verdict can be delivered. The same is true of geographical decision making — some of the sources used to reach conclusions may be irrelevant, inaccurate or even misleading. Therefore, the most important skill of the decision maker is the ability to *evaluate* and to identify *bias*.

Working with resources

Before beginning the analysis of resources, some key questions have to be addressed (Table 18).

Table 18

Question	Detail
Is the information relevant?	• Information selected must be relevant to the enquiry and focus on the issue or decision
Is the information valid?	• Check the source and the original purpose • Check that it is up to date and reliable — data from the internet are often up to date but may be less trustworthy than information printed in books or articles • Check who the authors are and who paid or commissioned them!
What are the advantages and disadvantages of the information?	• An advantage may be that the data are already processed and in a usable form • A disadvantage may be that there are too many data or the data may be too complex to be of practical use
Is there any bias?	There could be bias in: • the data supplied (or not supplied) • the way in which the data were collected (sampling bias) • the techniques chosen to represent and display the information
What are the limitations of the information?	• Data could be missing (e.g. people's/users' views) • There could be time periods when data have been omitted • It may be necessary to corroborate or substantiate the information from another source

Written resources

Written resources include books, newspapers, internet articles, magazine articles, specialist publications, government reports, data sets and so on.

Written resources may be produced *privately* or written *officially*. The latter are likely to be more reliable because official bodies have strict rules for collecting information, which may be enforceable by law, and they have resources to carry out research properly. Private sources are often biased (especially in their source information), may not be authentic and are not always produced by experts.

Evaluating written resources invariably involves some element of text analysis. Some ideas you might consider are:

- producing a written summary. Start by highlighting or underlining the most important aspects — this can be done as a group exercise — then swap information. Limit the rewrite to a particular number of words. This will make it easier to use in the exam.
- making a list of the key geographical issues raised in the text. This could be a series of bullet points or a table, with each point supported by additional text and figures.
- converting the text into a simplified 'mind map'. Mind maps are more compact than conventional notes, often taking up just one side of paper. Thoughts and factors are connected by lines, helping associations to be easily made.
- using a table to classify the issues into social, economic and environmental (or similar).

Example 1

On 3 November 2000, a local newspaper published an article about the 10 mph speed limit that is to be imposed on Windermere in 2005.

Summary

The 10 mph speed limit is certain to be introduced in 2005. In a meeting between the Lake District National Park Authority and power boaters, a claim was made that there will be loss of employment. However, other examples of imposing such a speed limit show no job losses occurring. Conservation (within the Park) must prevail over enjoyment — the area is 'not a playground'. Claims about loss of income from Windermere have been countered by the strong pound and high fuel duty.

Key geographical issues

- Conflict between lake users and the National Park Authority
- Potential loss of both incomes and employment because of the 10 mph speed restriction
- Possible fall in local tourism
- Conservation is more important than enjoyment within the Park
- The issue is complex, with little supporting evidence for either side

Example 2

The web provides information from all over the world. Because so much information is available, much of which is 'anonymous', it is necessary to develop skills to evaluate what you find. The internet has no filters — anyone can write a web page. Documents of a wide range of quality, written by authors with a wide range of authority, are available. Excellent resources are found alongside the most dubious. The internet epitomises the concept of *caveat lector*: let the reader beware. Nevertheless, you may use internet resources to help you understand a particular issue and they may be included as part of the examination.

When evaluating web resources, pay particular attention to the authentication of the author, the publishing body and any element of bias.

An extract from the 'No to ban on Windermere' website is shown below:

> In short, there are two bodies that want the ban: the South Lakeland District Council (SLDC) — an elected body that voted by a majority of one in favour of the ban — and the Lake District National Park Authority (LDNPA), a non-elected body (a quango). Both the SLDC and the LDNPA comprise salaried personnel that they deploy to ban our heritage and fun. Numerous voluntary organisations, lake users, commercial groups, chambers of trade, British Water-Ski, Windermere Water-Ski, classic motor club members, annual records week supporters, the Royal Yachting Association, Sport England and others are trying to negotiate for our rights. But volunteers do not have the resources of the SLDC and the LDNPA (which are ultimately paid by all of us/the taxpayers). Further to this, they deploy their resources and muscle to try to silence those who speak out, threatening free speech, leaflet distributions against the ban and taking trespass actions against peaceful protests on council property. Businesses are running scared that planning applications for those that object to the ban could be delayed (there are a ridiculous four tiers of planning approval in the LDNP). They are forced to comply because no sensible business can plan on a future where the ban may be overturned, and so it appears that they support the ban because of the changes they have to make. SLDC employees are scared; they dare not speak out against the ban and have even been prohibited from speaking at public enquiries. Democracy cannot be silenced.

Evaluating the resource:
- The author is an amateur water-skier.
- The extract is published on the author's own website, although there are links to registered bodies, such as the British Water-Ski Association.
- The extract is unbalanced and does not present any of the ecological or safety aspects that have brought about the proposed by-law.

How might a group in favour of the proposed speed limit react to the last argument?

Photographs and aerial-satellite images

The most essential skill with this type of resource is making use of the information that is displayed. You could start with a simple description.
- Start by 'unpacking' the general patterns, then continue by describing the more specific features or characteristics.
- It is often useful to work from the obvious known features to the unknown elements — go for the easy items first!
- If there are uncertainties, qualify your decisions by stating several possible options, giving reasons.

In an exam, you may annotate any photos or images provided. Be sure to make sensible extended remarks or comments, making use of the correct geographical terminology. Such annotations give more detail and show greater depth of knowledge and understanding than simple 'labels'. Any annotations you make can be included as part of your written answer.

Tip When describing a photographic resource, try to be as specific as possible. Wherever you can, use figures (e.g. size, scale, volume or speed) to strengthen your comments and remarks. It may a good idea to work in a group and share ideas about labels, annotations and comments etc. using pre-release resources.

Example 1

The photograph shows an urban street scene in Salford, Greater Manchester.

In this instance, the picture has been 'unpacked' by a group of geography students.

This is a zone of urban land use, showing some retail, residential and transport functions. At the time the photo was taken there was relatively low traffic flow and pedestrian density, suggesting the image was taken either away from the core CBD area or at a time (e.g. early morning) when there were few people around. The residential dwellings are typical of 1960s 'upwards' build (tower blocks), enabling high-density housing in areas of high bid-rent. Nowadays, this type of accommodation is often viewed as less desirable and is sometimes associated with a higher than average incidence of crime. The retail outlets look to be of a similar age, perhaps 1970, with a typical mix of concrete and blockwork, giving a utilitarian landscape. There is evidence of modernisation in the scene with the addition of flags and other steelwork, together with more recent fascia boards.

- The analysis starts with a general description.
- There is some interpretation to account for the low traffic flow and numbers of people.
- More detailed analysis is provided later in the section, for example dates of buildings and explanations.
- Some value judgements are used to link the scene with contemporary issues, such as crime.
- The section concludes with a more positive note about modernisation.

Example 2

A photograph of a mountainous area in mid-Wales has been annotated in a more traditional way, using arrows to indicate processes, features and explanations.

Steep-sided, catena-profile valley, showing modification due to recent glaciation

Mountain summits rise to nearly 600 m above sea level, giving rise to a harsh climate, >1200 mm rain per annum

Topography of the landscape suggests a hard, erosion-resistant rock (most likely igneous)

Mountain path (popular with walkers) provides ascent route to summit; the relatively gentle gradient (5–10%) improves route quality

Soils probably of relatively low quality (waterlogged in winter), and could be acidic in nature due to the high rainfall and acidic parent material

Small stream at the base of the valley is barely visible, although this stream would react quickly to rainfall events and have a 'flashy' hydrographic response

Valley sides are steep (30–40%), covered with low-quality, unimproved pasture; evidence of gullies (some with deep incision)

Very little evidence of settlement, although a road and field boundaries are evident in the far right of the photograph (western end)

Aerial/satellite images

The same rules apply when using these types of resource. Start with a description of what you see and then attempt an explanation. With aerial photos in particular, it is a good idea to include some idea of scale and size. This will be a 'best guess'.

Example

Below is an aerial photograph of the Greenwich Peninsula, London.

Here are some pointers for interpreting this aerial photo:
- First, orientate the image using a map.
- What evidence is there to suggest this photo was taken in the last few years?
- Can you use the scale of the river (or other features) to give an approximate scale for the map?
- Identify the dominant land use in the area (describe the distribution and form).
- Comment on the shape and flow direction of the Thames. Why is the river such an important feature in terms of economy and amenity?
- **www.multimap.com** is a useful site for practising these skills.

Maps

Maps are two-dimensional representations of the world, or part of its surface, as seen from above, in plan view. Originally produced for military purposes, maps now have a wider usage. They are important geographical information systems in themselves, conveying a rich stream of secondary data. Many issues analysis questions use maps, particularly Ordnance Survey (OS), to describe the location of the issue or decision.

Tip When you first handle a map, start by consulting the marginal information or the key. This should give the area covered, the scale, the orientation and the date of survey or publication. Maps are particularly useful in issues analysis, for instance in bypass options, or for understanding transport routes. Maps also provide a key to the past, giving us documentary evidence of old towns and villages and how they have changed over time. Maps help us to understand why certain location decisions might have been made; for example, the siting of an iron and steel works (near towns for labour and a port for transport). Remember that a good working knowledge of maps will not only help you in Unit 6, but also in other aspects (e.g. coursework).

Types of map
- **OS maps.** These have a variety of scales and levels of detail and are the most likely type of map to be used in issues analysis.
- **Historical maps.** Early historical maps date to around the sixteenth century, but historical maps can be recent, such as those showing changes in town centres (Goad maps).
- **Sketch and base maps.** These may be hand-drawn and often simplistic outlines. Contour lines are usually omitted. They show selected details, depending on the purpose of the map and the cartographer.
- **Road maps.** These are designed specifically to show transport routes and networks (they are often enlarged and not to scale). Often, places of interest, such as museums, are shown, but other details are omitted.
- **Specialist maps.** These vary and can include town centre maps, maps on how to get to particular locations (often produced by organisations), geology maps and land-use maps.

Using maps
- First look for general trends, such as the distribution of particular features.
- Use the correct terminology to describe the degree of concentration (regular, random or clustered), using supporting location information.
- Look at the shapes of particular features, especially settlement patterns — nucleated, linear or cruciform.
- When referring to places or features, be specific. Use scales to calculate distances and grid references to locate points. The latter can either be four-figure for a 1 km square or six-figure for a more specific, smaller feature, such as a sports centre. Support grid references with place names.
- Use the orientation of the map to describe the trends of any features (e.g. 'the river valley runs NNE to SSW'). Appreciation of gradient may be important.
- Certain maps may reveal anomalies. Locate these precisely, and state how they vary from the trend or average.
- Maps (including sketch maps) may be annotated neatly, with lines pointing to the appropriate location.

Example
Overleaf is an extract from an OS map (1:50 000). Note the style of annotation and the precision of the descriptors for particular features.

SHREWSBURY

In the SE quadrant of the map, there are large areas of parkland, occupying in excess of 1 km²

Grid square 5413 is dominated by a coniferous plantation, which rises to a height of 153 m

The main rail link shows an WNW–ESE trend; the dual carriageway also follows this trend from square 5212

The area to the SW of Preston (529118) is low-lying and probably flood-prone

There is a range of residential layouts: 5012 is a tradional grid pattern (likely Victorian terrace), whereas 5113 appears to have a more modern estate layout, indicated by the cul-de-sac pattern

The original site of the old town is in a meander loop, defended by a castle — a dry point above level of flooding

On the inner side of the dual carriageway bypass, the roads exhibit a radial pattern, converging on the centre of the town

Approximately 16 km² of the map is built-up area; the remainder is a mixture of agricultural land-use, woodland and small hamlets/villages

Grid square 4909 has a large quarry, running NE–SW and linear in form; prevailing winds from the SW may mean that Betton Strange (509093) and Weeping Cross (514103) would suffer from noise and dust pollution

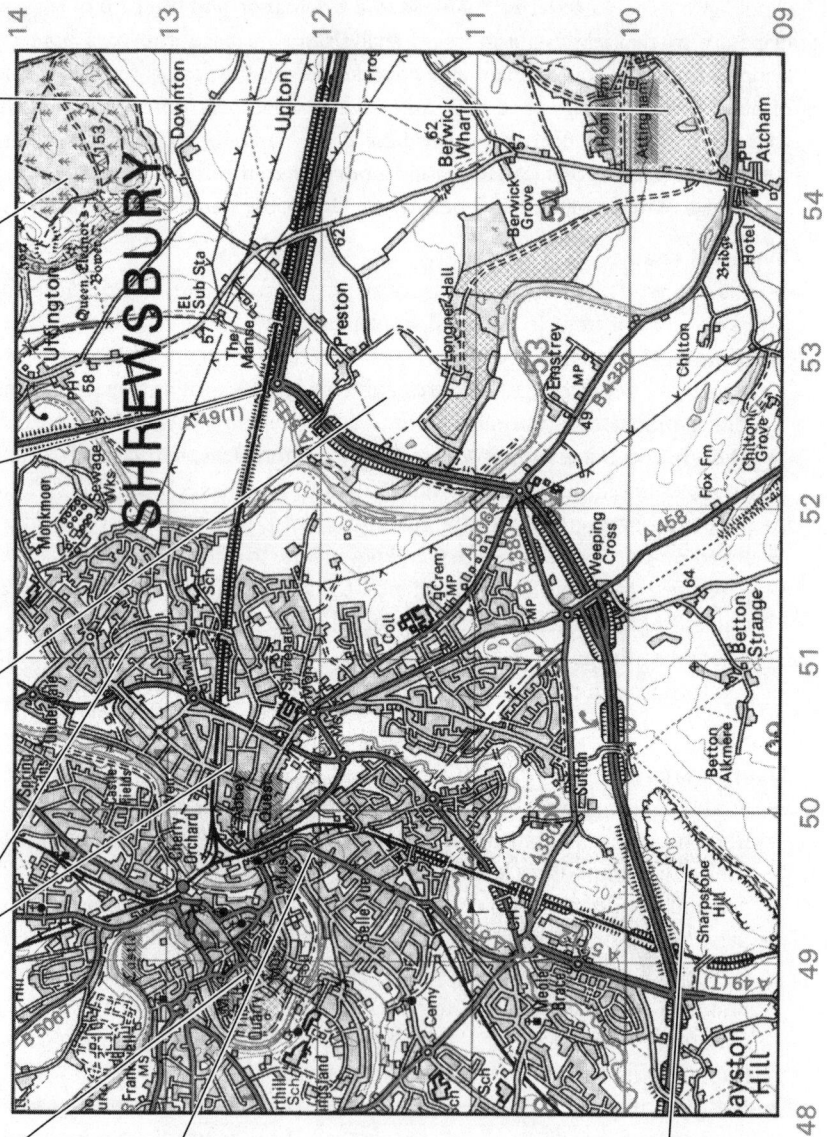

Graphs

There are many different types of graph. To prepare for the exam you should familiarise yourself with the data representation techniques used in the pre-release materials.

Graphs have a number of purposes:
- The main purpose of a graph is to display data in a way that is easy for someone else to understand and interpret.
- Graphs summarise data.
- Graphs may demonstrate points more forcibly than when the data are presented as text or tables.

Using graphs
- Start by describing the general trends. Use descriptors such as up/down, decrease/increase or steep/gentle. Account for any anomalies.
- If the graph is complex and no obvious overall trends are indicated, split the graph into manageable parts (phases or stages). Describe each stage in turn. Again use rates of change and values. Annotate the key features on the graph itself.
- Identify significant changes and offer realistic explanations.
- Describe relationships precisely — e.g. positive/negative correlation, inverse relationship and exponential trends.

Tip It is a good idea when describing graphs to use precise language and support it with figures — for example, 'there is a gentle increase of 20% between time *x* and time *y*'. Percentage change can be estimated by eye or calculated using the following:

$$\frac{\text{time } x \text{ data} - \text{time } y \text{ data}}{\text{time } x \text{ data}}$$

If time *x* data = 22 and time *y* data = 15, then percentage decrease = $\frac{22 - 15}{22}$ = 32% drop.

Example
Figure 13 shows the UK population age profile (pyramid) for 2001 compared with 1951.

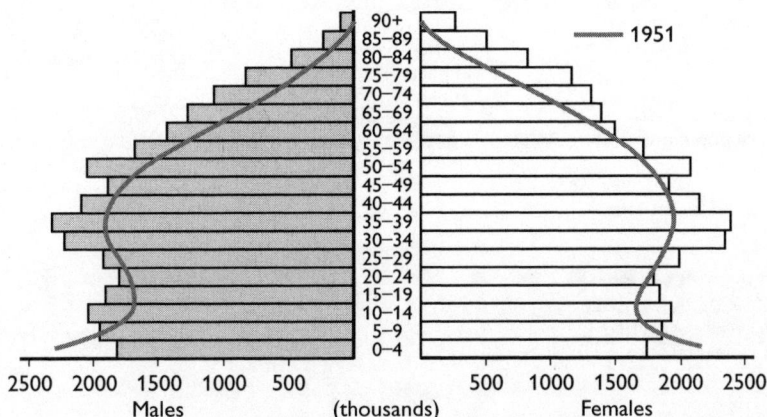

Figure 13

If you had to describe the data, you might write something like this:

- Between 1951 and 2001, the population increased in nearly all age groups.
- The graph shows that there are still more women living longer than men. In 2001, there were 4.1 million men aged 65 and over compared with 5.3 million women.
- In 1951, there were around 0.2 million people aged 85 and over (0.4% of the total population). In 2001, this had grown to just over 1.1 million (1.9% of the total population).
- Only the 0–4 yr category shows a decline over the last 50 years.
- Overall, the proportion of the population aged under 16 has decreased to 20%, compared with about 24% in 1951.
- There is a significant increase for both males and females in the modal age category (35–39 years) — a rise from 1.8 million to 2.3 million.
- There is also a significant increase (60%) in the 75–79 age group.

Diagrams and tables

Diagrams take various forms, so expect some variety within the issues analysis exam. A diagram may be a qualitative representation of a theory, concept or process. When analysing diagrams, apply the same rules that you would for graphs and photos — be specific and precise. Be prepared to make a list of advantages and disadvantages of the illustration.

Figure 14 is an extract from a diagram to show the negative impacts of the proposed expansion of Dibden Bay port in Southampton.

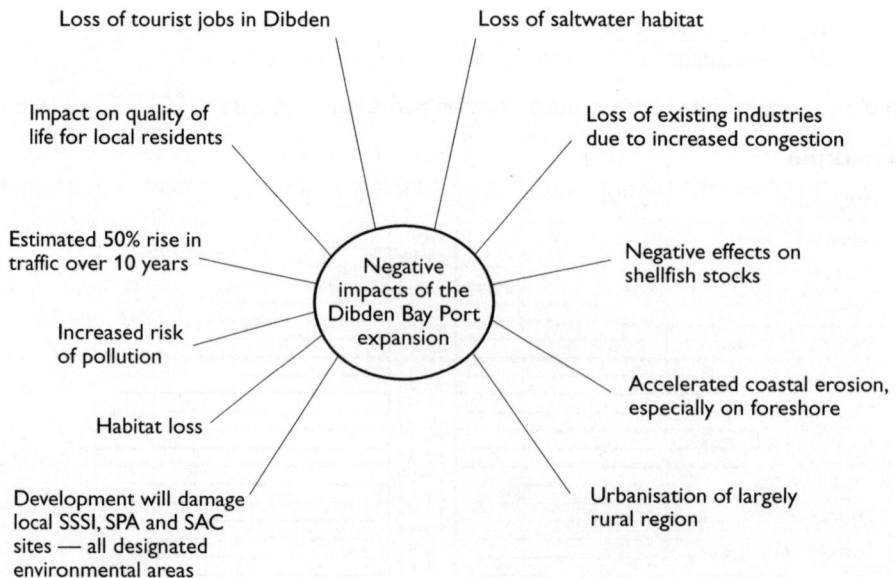

Loss of tourist jobs in Dibden

Loss of saltwater habitat

Impact on quality of life for local residents

Loss of existing industries due to increased congestion

Estimated 50% rise in traffic over 10 years

Negative impacts of the Dibden Bay Port expansion

Negative effects on shellfish stocks

Increased risk of pollution

Habitat loss

Accelerated coastal erosion, especially on foreshore

Development will damage local SSSI, SPA and SAC sites — all designated environmental areas

Urbanisation of largely rural region

Figure 14

Advantages and disadvantages of this diagram are summarised in Table 19.

Table 19

Advantages	Disadvantages
Considers a wide range of socioeconomic and environmental factors	Only shows the negative impacts, and is therefore unbalanced (positive impacts include increased employment opportunities)
Some descriptions contain additional details (including supporting figures), giving justifications	Not all points are validated with supporting information
Easy to distil pertinent facts from the range provided	Shows nothing about the relative strengths (ratings) of the disadvantages stated

Questionnaire data

Questionnaires are used to obtain information from specific groups of people. They are usually forms with sets of pre-arranged questions, which assess the target groups. Information gathered might include:

- basic sociological characteristics — e.g. age, gender, occupation or religion
- spatial patterns — e.g. origin of visitors or typical travel routes on a journey to work
- patterns of behaviour — shopping habits or motives/preferences for recreation

In an exam context, you are likely to be presented with partially or fully processed data. You should pick out major trends and identify possible anomalies, which you can use to support statements you make for or against a particular issue or decision.

In the example below, 100 people were surveyed in a honeypot location. Their reasons for visiting are identified according to 13 categories (see Table 20).

Table 20

Age (years)	<21	21-35	36-50	51-65	>65
Dog walking	I	I	IIII	IIII	II
Visiting friends			III	II	
Running/jogging	I	I	I		
Sightseeing		I	卌	卌 II	卌
Walking (>2 miles)		II 卌 卌	卌 II	IIII	II
Picnicking	II		II		
Mountain biking	卌 III		I		
Use facilities	II		I	IIII	卌
Take a rest	I				
Education	卌 III				
For the children			II		
Nostalgia		I			
Other (specified)					
Totals	23	15	27	21	14

- The modal age group of respondents interviewed is 36–50 years — 27%. The least represented age group, only 14%, is people over the age of 65.
- In the youngest age category (<21 years), education and mountain biking are the most popular activities, with around 70% of the responses.
- 'Nostalgia' and 'Take a rest' are the least popular activities, accounting for only 2% of the responses.
- Most respondents (80%) in the age category 21–35 years visited the location to go walking for a distance of more than 2 miles. This is the most popular activity across the total age range.

Interview data

Interviews are different from questionnaires, in that the questions and responses tend to be longer, less structured, more detailed and open-ended. Often, the results of an interview are presented as a transcript (a direct written copy of what the respondent said). The first thing to do is to make sense of the responses.

- Read through the commentary or transcript and use a highlighter pen to 'extract' the significant elements. You should highlight a maximum of 20% of the material.
- Alternatively, use two different coloured highlighters to pick out comments that are in favour and those that are against. These could then be tabulated.
- Use 'talking heads' or summary boxes to show relevant information from a particular interview. Summary boxes should contain a maximum of three sentences.

Cartoons

These may be difficult to interpret. It is often best to brainstorm ideas as a group.

**Figure 15
Battling for
Britain's beaches**

For the example shown (Figure 15):

- identify the 'parties' (groups of people) who are 'battling' for Britain's beaches
- develop a conflict matrix to see how tourism, fishing, coastal development, shipping and military use have the potential to conflict with each other
- develop your synoptic skills by thinking of similar examples of coastal damage

Evaluating options

Options analysis is at the heart of many decision-making exercises. It is about giving well explained, well argued reasons and being able to see the 'bigger picture'. Lines of questioning test your ability to appraise and assess schemes and sites:

- Justify your decision to locate at W.
- Outline the choices available to manage the problem at X.
- What are the advantages and disadvantages of site Y?
- Give alternative locations for the proposed development at Z.

Evaluation of options tends to be either assessment of various schemes or choosing the best location for a development.

Assessment of schemes

Figure 16 is adapted from a publicity poster produced by the Environment Agency in 1993 to inform the residents of Shrewsbury about the flood option schemes that had been put forward.

Figure 16

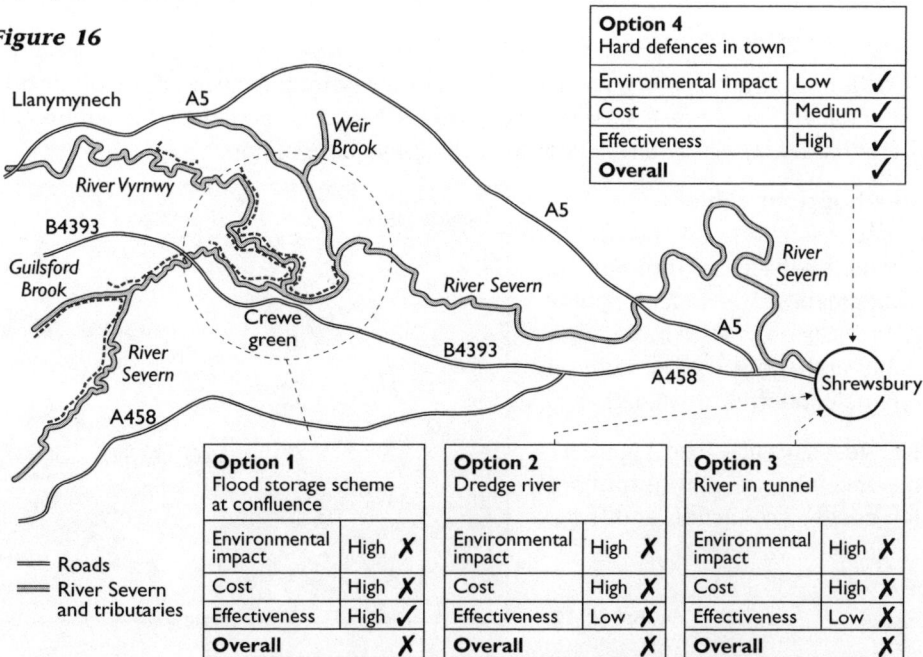

Option 4
Hard defences in town

Environmental impact	Low	✓
Cost	Medium	✓
Effectiveness	High	✓
Overall		✓

Option 1
Flood storage scheme at confluence

Environmental impact	High	✗
Cost	High	✗
Effectiveness	High	✓
Overall		✗

Option 2
Dredge river

Environmental impact	High	✗
Cost	High	✗
Effectiveness	Low	✗
Overall		✗

Option 3
River in tunnel

Environmental impact	High	✗
Cost	High	✗
Effectiveness	Low	✗
Overall		✗

The options are analysed in Table 21.

Table 21

Option	Description	Outcome
1	Flood storage scheme	Effective, but costs and environmental impact too high
2	Dredge river	Rejected — too expensive, too great an environmental impact, together with low effectiveness
3	River in tunnel	Rejected — too expensive, too great an environmental impact, together with low effectiveness
4	Hard defences in town	Preferred option — low environmental impact, moderate cost and high effectiveness

There are a number of drawbacks in the way the information was presented:

- Four options do not cover the entire range of possible schemes. For example, the river could be diverted upstream or additional flood storage schemes might be implemented.
- There is a limited range of 'indicators' for each option. The public information sheet does not take into account the views of local residents, particularly in terms of aesthetic disturbance.
- Each indicator is qualitative, with no indication of monetary value. What is the differential between low, medium and high cost? On what scale is environmental impact calculated?
- The flooding problem in Shrewsbury might be tackled through the use of a range of approaches, such as a mixture of hard defences (walls) and upstream management procedures.

So, options analysis is complex as there is a range of intangibles that are difficult to cost and identify. Option 4 was decided to be the best scheme. Why do you think this was? What additional criteria might be used to evaluate options?

Use of options tables

A more systems-based approach to options analysis is through the development of a series of 'options tables'. This provides a more quantifiable end result and is therefore preferable in terms of validation.

The Western Isles (see Figure 17) face special problems in terms of high costs of collection and transportation of waste.

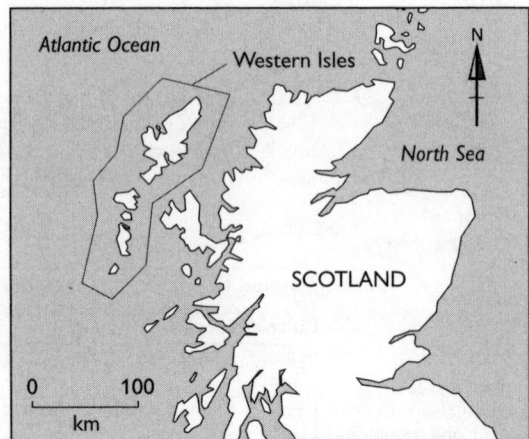

Figure 17

The Waste Management Plan provides a framework to facilitate strategic planning for the next 20 years. Five options have been identified as part of the future plan for the disposal of waste:

- Option 1 — continued use of local landfill sites
- Option 2 — development of a new energy-from-waste incinerator to produce power
- Option 3 — development of a new, centralised composting plant
- Option 4 — development of a new, centralised anaerobic digestion plant
- Option 5 — export of waste to the Scottish mainland

Table 22 is an options table for waste management in the Western Isles.

Table 22

	Option 1	Option 2	Option 3	Option 4	Option 5
Waste minimisation	Encourages local initiatives	Encourages local initiatives	Encourages local initiatives	Encourages local initiatives	Encourages local initiatives
Recycling and reprocessing	Increased recycling of glass, paper, plastics and metals	Increased recycling of glass, paper, plastics and metals	In-vessel composting of all bio-degradable waste at central facility; increased recycling	Anaerobic digestion of all bio-degradable waste at central facility; increased recycling	Increased recycling of glass, paper, plastics and metals
Recovery	None	Energy produced	Soil improver	Energy produced; liquid fertiliser and soil improver	None
Disposal	Existing waste quantities to landfill	Reduced quantities to landfill (ash)	Significantly less waste to landfill	Significantly less waste to landfill	No disposal in strategy area
Capital cost (£m)	4–6 Council-funded	7–10 Probably grant-assisted	4–6 Probably grant-assisted	4–6 Probably grant-assisted	2–3
Disposal cost (£ per tonne)	70–80	50–80	30–40	30–40	100–120

Depending on the question and the nature of the issue analysis, the next task is to select suitable criteria on which justified decisions can be made (Table 23).

Tip In the unit test, there is usually guidance provided in the letter of instruction to help you establish suitable criteria.

Table 23

Criterion	Details and description
Environmental	What are the impacts on air, land, water, natural and cultural heritage, global warming, local amenity and non-renewable resource use?
Economic	Does the option bring money into the Western Isles, creating new markets and commercial opportunities? This includes savings from recycling and resource re-use (e.g. providing local employment opportunities).
Social	Implications for the population in terms of jobs and opportunities to develop local skills. Will the options achieve the aspirations of the local communities (e.g. public acceptability, making people more responsible for waste)?
Practicability	Would the option work in the Western Isles? In particular, in terms of technical feasibility, flexibility to meet future technologies and use of existing facilities.
Compliance	Will the proposed option fall in line with local, national and European legislation, such as the Landfill Directive and renewable energy policy?

One way of evaluating the options is to retabulate them, using the template above. All the details for each option have to be explored. In Table 24, Option 1 has been completed as an example.

Table 24

Option 1: landfill (no change)	
Criterion	**Details and description**
Environmental	• Poor use of reusable and recyclable materials • No energy recovery • Difficult to control emissions to both air and water • Any new landfill cells would have to be built to a higher environmental standard
Economic	• Landfill tax rises in the future could increase the tax bill from the current £350 000 to £750 000 (annually) • From 2010, permits will have to be bought, in accordance with the Landfill Directive
Social	• The system is already in operation • Unlikely to provide further employment or new skill opportunities • Encourages very little 'ownership' or personal responsibility for waste
Practicability	• All the infrastructure is in place • New waste transfer stations will have to be built to store waste for transportation and disposal at a central site
Compliance	• Very poor compliance with other policies • Waste minimisation and local recycling initiatives will reduce the amount of organic waste for landfill, but this will probably not be enough to reach Landfill Directive targets

Best practicable environmental option (BPEO) is a useful idea. It describes the most effective long-term solution from a list of options. BPEO decision making is systematic and consultative, based on a comprehensive assessment of the environmental, social and economic impacts of alternative options. The BPEO for the Western Isles has been assessed as either option 3 or 4, or a combination of both, with a preference for anaerobic digestion. This has the benefit of producing a valuable local energy supply. Both options:

- would provide an effective means of meeting local requirements
- would generate a valuable product for improving local land
- would benefit the local community and economy
- have the potential to assist local industry sectors to recover more waste
- received strong support from the local community

Choosing the best location for a development

Another aspect of options analysis is evaluating the best location for a new development from a number of choices. You may be asked to examine the nature of sites within a small geographical area, such as within a village or small town. You might have to carry out options analysis for a wider geographical area, such as a region or country. Ideas that might be relevant in site selection include:

- the physical nature of the site — shape, size, topography and drainage
- site ownership — are the current owners willing to sell and how quickly can the land be released?
- cost — purchase and development
- potential assistance in development — may be relevant in an area targeted for regeneration
- surrounding land use — congruence is important
- site accessibility — closeness to motorways and other main transport routes
- planning restrictions on the site — especially if it is greenbelt
- greenfield/brownfield

Choosing the site for the new National Stadium

The new National Stadium is a government-backed project which has highlighted the difficulties in choosing a site. Three sites were proposed:

- the same location as the old Wembley Stadium in London
- a Birmingham-based location near to the NEC in the West Midlands
- a Coventry-based location at Foleshill, to the north of the city (see Figure 18)

Figure 18

Table 25 sets out some of the considerations in the site options analysis. The costs, based on a 2001 scoping report, are approximate.

Table 25

	Birmingham	Coventry	Wembley
Site status	Greenbelt; local pressure groups have expressed concern	Planning permission exists for a 40 000-seat stadium	New permission required for larger stadium; requires listed building demolition consent
Planning risk	High	Low to medium	Low to medium
Site size (acres)	125	73	47
Site ownership	Birmingham City Council	Coventry City Holdings Ltd	Wembley London Ltd
Site availability	Immediate, subject to planning consent	Requires negotiation with Coventry City FC	Immediate, following demolition
Rail (people/per hr)	10 000	10 000	30 000–50 000
Road, rail and air access	J6, M42; Birmingham International Airport	J3, M6; rail 6 km away	Wembley Park tube (requires upgrade)
New infrastructure cost (£m)	30	40	120
Car park spaces	23 000 including NEC	4900	3200
Total project cost (£m)	495	470	750
Funding gap (£m)	80	160	200

This is a particularly complex options analysis, not least because of the strong feelings held by the football fans in various areas. In 2001, a survey carried out by the University of Leicester, involving 2000 fans from 43 clubs, strongly supported relocation to Birmingham. The leader of Coventry City Council petitioned both the Football Association and the Secretary of State for Culture, Media and Sports in support of the Coventry bid. However, one of the main factors that provided Wembley with success was the site's ability to handle large numbers of visitors. The predicted visitor flow capacity (people per hour) is shown in Figure 19.

A site options analysis could take a number of forms:
• tabulating advantages and disadvantages for each site and location
• ranking specific elements of data
• retabulation of data, including descriptive prose

Only three site options were serious contenders for the new National Stadium. Use a small-scale map of the UK to propose alternatives. Below are some ideas that you might find useful.

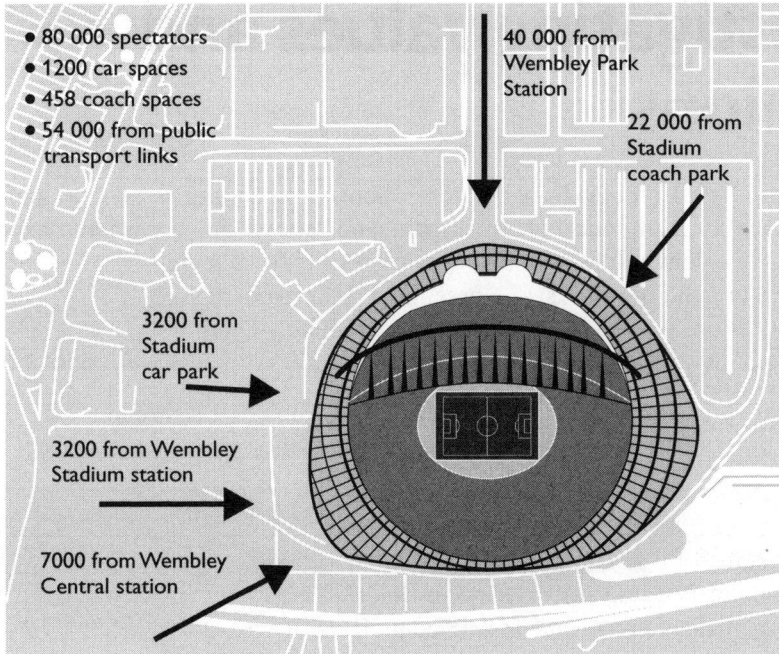

- 80 000 spectators
- 1200 car spaces
- 458 coach spaces
- 54 000 from public transport links

40 000 from Wembley Park Station

22 000 from Stadium coach park

3200 from Stadium car park

3200 from Wembley Stadium station

7000 from Wembley Central station

Figure 19

- Where are the traditional heartlands of football? For example, there is a great passion for football in Liverpool. Both Liverpool and Everton have recently announced plans for new stadia.
- The northeast of the country seems to be under-represented in terms of proposals/bids. What are the potential opportunities in this area, particularly around Newcastle and Sunderland?
- The southwest and Bristol are now very accessible by road, rail and air. Are there big enough centres of population to warrant relocation here?

When a suitable geographical area has been found, you need to consider the specifics of the site, in particular:

- size — what is the minimum and maximum area? (Shape may also be important. Is it convenient for a stadium, i.e. square or rectangular versus long and thin?)
- type of land — brownfield sites are often expensive to clean up, but may attract grants. There is less planning risk associated with brownfield locations.
- population — how many people live within 50 or 100 miles? Try to construct isolines of population.
- access — list the ways of travelling to the site and the capacity of each method

Make a note of possible fieldwork opportunities that could be carried out to further validate, or invalidate, your choice.

Investigating values and opinions

Values are based on beliefs or opinions about an issue and result from upbringing, age, education, influence of the media and, above all, a person's experience. Particularly emotive or sensitive developments might include:

- where to build new houses — greenfield or brownfield sites?
- the location of new waste facilities — particularly landfill
- siting of new refugee accommodation — in towns, in the countryside or not at all

Since there is such a variety of interested parties who often contribute to the decision-making process, the first task is to make sense of the opinions expressed.

The example chosen is the Dinas Dinlle Coastal Works, North Wales (see Figure 20). The coastal front of Dinas Dinlle has been susceptible to flooding since Victorian times, due to poor land drainage and tidal inundation.

In 1994, a plan was put forward to construct two rock promontories to protect the village front, supported by beach management, thus allowing the shore to revert to its natural position.

The views expressed about the proposed development are summarised in Table 26.

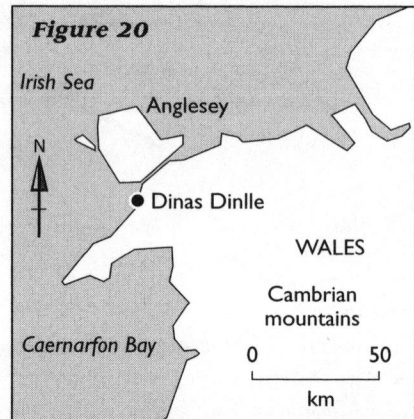

Figure 20

Table 26

Interested party	View expressed
Countryside Council for Wales	It is difficult to predict the impact of the structures on the landscape. They represent an artificial intrusion along a relatively undeveloped coastline.
Arfon Borough Council	The plan will need to reconcile the potential for further exploitation of tourism with the ever increasing importance of environmental issues.
Dinas Dinlle resident	The beach is an important resource used by visitors, who bring income into this area of Wales. I hope these developments will not undermine the aesthetic quality of the area, but I recognise the need for coastal protection.
Sea Fisheries Committee	Beach netters will be most affected, but they are mobile and can work to the north of the area. We cannot see the developments having a detrimental effect on foreshore fisheries.

Interested party	View expressed
British Trust for Ornithology	Loss of habitat at Morfa Dinlle could seriously affect the total number of birds using this part of North West Wales. The area is an important feeding ground for golden plover and shelduck.
Gwynedd Planning Department	The proposed works might only have a minor aesthetical impact on the view of the Heritage Coast, which is 2–3 miles to the south.
Environment Agency	Our partnership with Arfon Borough Council is to allow the sea to erode part of the land, which includes the current minor (B) road, over a long period of time (20 years or more) to create a new coastline. This will secure the future of Dinas Dinlle village.
Lecturer, Bangor University	We cannot afford to defend this section of Welsh coast. It is simply not economically sustainable. Managed retreat must be encouraged — it's too hard to battle against the sea.
Caravan Park owner, Dinas Dinlle	We can only open our site to campers during the driest parts of the year, as the surrounding land here is so wet. The NRA's plans should help drain and dry out our site so we can extend the season.

Table 27 below shows one way of analysing some of these views. You could extend the table to include all the views.

Table 27

Interested party	Summary of opinion	Status of interested party	Possible evidence to support view
Countryside Council for Wales	'On the fence'; recognises the impact, but also appreciates the need for managed retreat, combined with new hard defences	Official government body	Limited evidence, though maps and plans might support the views put forward
Dinas Dinlle resident	Wants the development to proceed, but concerned about environmental impact	Local person, emotionally involved with the decision	Data from visitor numbers, popular part of Welsh coast (Lleyn Peninsula); close to Snowdonia
British Trust for Ornithology	Development of the coastal works will have significant impact on bird populations and their environments	Official from a recognised charity, supporting environmental care for birds and their habitats	Data confirming site's importance as bird breeding ground
Lecturer, Bangor University	The area should be left alone with no implementation of coastal works	Academic	Research from other coastal areas

Evaluating opinions

There are a number of other ways in which opinions can be evaluated:

- Construct a matrix. This works well if you have a number of possible sites or alternative locations (see Figure 21).

Opinion \ Site	1	2	3	4	5
1					
2					
3					
4					
5					
6					
7					
8					
9					
Totals	6/9	8/9	3/9	4/9	3/9

Add this row to summarise and to add totals

Figure 21

In this example, shaded opinions represent people who consider the location most suitable for development. Site 2 has the highest number, followed by site 1, with sites 3, 4 and 5 scoring the lowest.

- Develop a composite table, using criteria specific to the exam question or purpose.

Monitoring techniques: environmental impact assessment

Environmental impact assessment (EIA) is the systematic analysis of the likely ecological and social consequences of a proposed project or development. It is a type of audit that takes into account future costs and benefits. In the UK, EIA is mandatory for large-scale projects.

EIA has its roots in the concept of sustainability and is seen as an essential tool to make good, rational decisions. Therefore, it is used as a planning tool. It is closely linked with CBA (see pp. 27-29) but is generally less quantitative and wider in scope.

Tip EIA is complex and in some instances EIA decisions are highly subjective. Be prepared to question the outcomes from a published EIA — no two individuals or the organisations they represent will view the environment in the same way.

How to carry out EIA

In the final part of the issues analysis examination, you may have to look at impacts of any decisions made. This section shows you how to carry out some post-decision analyses.

Most EIAs are based on interactions between the project characteristics and environmental factors. One of the first interaction matrices to be developed was the Leopold matrix.

The Leopold scoring system

The Leopold scoring system is a more conventional (and more complex) method of evaluating impacts. It involves splitting cells with diagonal slashes (see Figure 22).

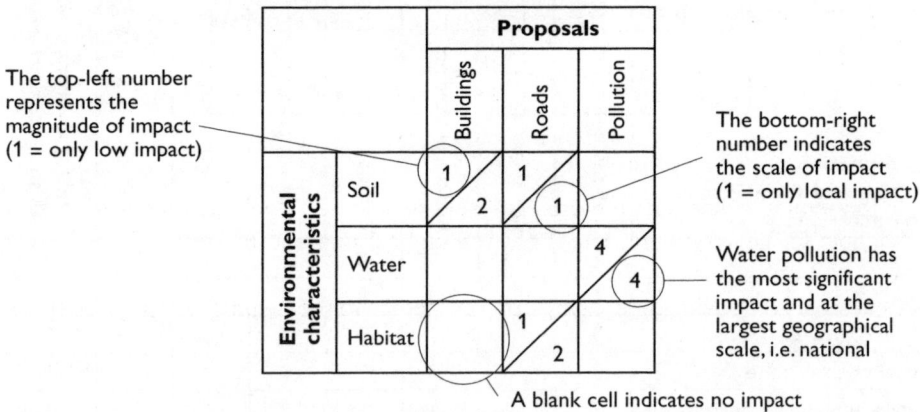

Figure 22

Tip You can apply a weighting to certain project characteristics to modify their significance in the overall decision.

Now, many types of matrix are used, mostly based on the Leopold scheme.
- Start by making a list of the activities, such as increased noise and traffic, that characterise the project. You may decide at this point that two matrices should be developed — one for the construction phase and one for the operational phase.
- Make lists of the environmental factors that will be affected and the social impacts that will occur.
- Arrange the lists in a grid or matrix, with project characteristics along the top and environmental factors down the side.

The way in which the impacts are evaluated opens up a number of options, each with varying degrees of difficulty.

A simple scoring system

- Take each project characteristic in turn and consider its impact on the environmental factors.
- Use a basic method to classify the impact:
  - Score 0 = no impact
  - Score 1 = low impact
  - Score 2 = moderate impact
  - Score 3 = high impact
- If the impact is difficult to categorise, use split values (e.g. low to moderate impact scores 1.5).

Table 28 is a basic impact matrix for the construction phase of a proposed reservoir in an area of scenic beauty.

Table 28

Project characteristics / Environmental factors and social characteristics	Traffic issues	Raw materials	Additional buildings	Workforce	Waste disposal	Reservoir/water supply	Landscaping	Total for environmental and social characteristics
Flora	1	2	1	0	1	1	3	9
Fauna	1	2	2	0	1	1	3	10
Hydrological characteristics	0	1	1	0	1	1	3	7
Local microclimate	0	0	1	0	0	0	0	1
Landscape and intrinsic value	2	2	1	1	2	1	3	12
Tourism resource	2	3	2	1	2	0	1.5	11.5
Noise	2.5	0	0	2	0	0	3	7.5
Air quality	0	1	0	0	0	0	1	2
Local residents	3	2	1	2	2	1	1	12
Local landowners	1	1	1	0	1	0	1	5
Total for project characteristics	12.5	14	10	6	10	5	19.5	

Tip When developing a matrix, try to keep things simple. Go for a maximum of 10 × 10 headings — this still gives 100 cells to evaluate!

Interpreting matrices

Any text accompanying the matrix should be a discussion of the most significant impacts. Options include:

- commenting on those columns and rows with the largest numbers (you might consider comparing these values against the mean, mode or median). In the simple scoring system example in Table 28, there is significant impact on the tourism resource and the landscape quality (totals of 11.5 and 12 respectively), while modification of local microclimate has the least impact.
- highlighting individual cells that might be anomalous, having particularly high or low values.
- for more complex EIAs and proposals, constructing your own summary checklist, highlighting some of the significant factors. Table 29 is an extract from a summary EIA checklist for the development of a new local bypass.

Table 29

Environmental factors	Social factors
• Environmental protection measures required for flora, fauna and scenery • Effects on soil (e.g. drainage, structure, quality) • Effects of noise and air pollution • Impacts on existing environmentally sensitive areas and protected areas or features (e.g. SSSIs) • Specific impacts from the constructional phase	• Proximity to residential areas (effects on people and communities) • Conflicts with planning policies at a range of scales • Changes in property values and land use • Safety of pedestrians and vehicles; hazard and emergency services effects • Implications/predictions of future traffic patterns and growth

EIA case study

A new wastewater treatment works (WWTW) is needed in Hull to comply with a European Directive, relating to the quality of sewage discharged into inland and coastal waters. Sewage from Hull and the surrounding areas is currently discharged directly into the Humber Estuary from two pumping stations. The plan is to divert sewage via a deep tunnel to a new WWTW located in east Hull (see Figure 23). The sewage will be treated to reduce its pollutant load prior to discharge.

Environmental assessment:
- **Effects on land use, amenity and recreation** — minor effects on land use and amenity will arise during construction and there will only be a small loss of land. With regard to recreation, there will be a temporary and limited loss of open space, including short-term diversion of a footpath. Planning associated with the WWTW will improve visual appearance of the adjoining footpath. There will be a general improvement in water quality — a benefit to recreational users.
- **Agriculture** — the transfer will not affect agricultural land, but the WWTW will be built on grassland that was taken out of production some years ago.
- **Landscape and aesthetics** — most of the scheme will be constructed in already built-up areas. There will be a temporary loss of landscape quality until the new

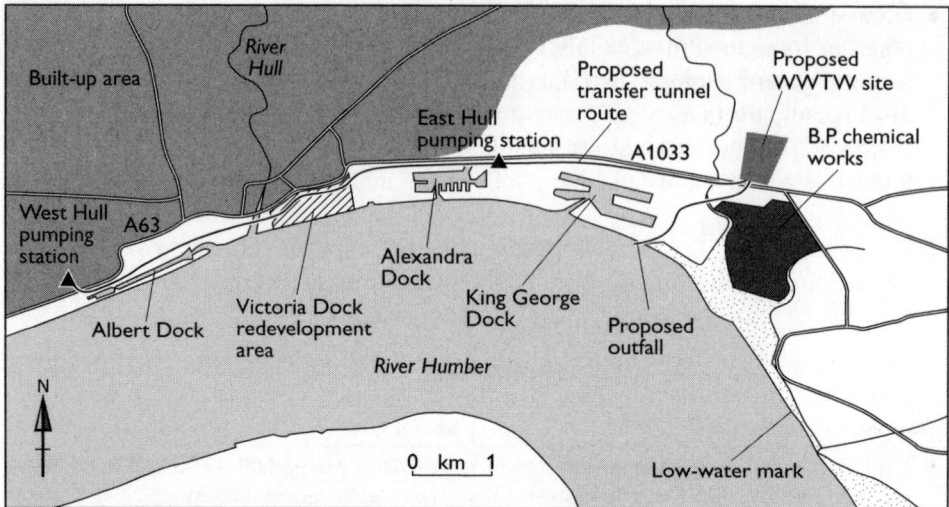

Figure 23

landscaping of the WWTW becomes established after 5–10 years. Some of the WWTW will still be visible after landscape maturity, but screening with trees will reduce this impact. Night-time lighting of the WWTW will be visible, but this will be minimised to reduce impact on local residences.

- **Socioeconomic factors** — potential community issues include negative perceptions about the scheme (disturbance and nuisance). In the short term, the scheme will create 330 temporary jobs, but only 15 permanent jobs.
- **Air quality and odour** — there is potential for dust during construction and odour during operation of the scheme. Dust nuisance can be minimised by control measures; odour levels have been computer modelled and will have a negligible effect on local properties.
- **Noise and vibration** — construction noise will create audible disturbance, especially to properties at access shafts along the transfer pipe. Properties near the WWTW will experience slight to moderate noise impact. Once operational, the WWTW noise disturbance will not affect local residents. Vibration impact will be noticeable during construction; low-impact piling methods will be used.
- **Water quality** — during construction, controls will be used to prevent pollution of the Humber Dock and Marina, the River Hull and Holderness. Computer modelling has shown that the WWTW plant will provide a general improvement in the estuarine water quality, although there will be an increase in nutrient levels. There will be some minor effects from storm-water discharges on the existing mudflats.
- **Ecology and nature conservation** — the Humber Estuary is important locally, nationally and internationally. During construction, there will be short-term loss of habitat and disturbance to waterfowl on the estuary. Operation of the scheme will alter the number and distribution of bottom-dwelling marine life. However, this is countered by improved water quality.

- **Construction waste and disposal** — demolition and construction waste will be taken to local landfill sites. There will be minor impact during waste handling, transport and disposal.
- **Traffic and safety** — slight impact will occur locally during construction, from temporary traffic lights at both shaft sites and at the WWTW. No longer-term impacts are envisaged.

Activities suggested

- Develop an EIA matrix using the brief details provided in the project summary. Use the headers in the above summary as a way of listing the 'environmental factors'. You will have to define your own project characteristics.
- You may want to split the EIA into two separate matrices: one to consider the impacts of the treatment works; the other evaluating the effect of the new transfer pipe.
- Sketch the map provided on a sheet of A4 paper and suggest alternative locations for the WWTW site. Mark and rank (from most to least favourable) possible alternatives on your map, annotating with justifications.
- Annotate the map with some of the most important environmental impacts that are identified in the environmental statement.
- Describe the route of the proposed sewage and storm-water tunnel. Suggest reasons why the transfer link has been planned to take this route.
- Suggest alternative and sustainable land uses for the redundant East and West Hull pumping stations.

Tip You will probably not have time to use all the above analytical techniques in examination conditions. However, you need to be aware of the wide range of techniques available to help you summarise, synthesise and analyse resources.

Handling pre-release resources

The resources are released to centres about 3–4 weeks before the January exam and six weeks or more before the June exam. There are essentially four major tasks to complete in your preparation:

- Researching the *geographical background* behind the issue and thinking how you could demonstrate synopticity.
- Analysing the full range of *resources* — in particular, concentrating on getting the most out of maps and photographs, some of which are very complex. Text needs analysing and highlighting in detail (see p. 37).
- Exploring the *opinions* about the issue (values analysis) — (see p. 56).
- Evaluating the *options* on offer to understand the likely decision. This may be a choice of sites, a choice of schemes, relating schemes to sites, or a choice of options (see p. 49).

Always remember that you do not know the precise questions, so you need to be on a fact-finding mission, looking at all angles. The end product should be a set of well-researched notes, resulting from data analysis, synthesised in a form that is easy to assimilate (best done the night before) with diagrams and lists to help you remember the key facts.

Preparation programme

The programme outlined below assumes the recommended 10 hours (combining class and individual preparation time). Time allocations are approximate.

The resources arrive
Have a really good read through them to try to figure out what they are about. You may have the opportunity to work through them in class, discussing photos, interpreting statistics and graphs and analysing maps. (Allow 2 hours.)

Key classroom sessions
Be there! Follow guidance from the teacher on what geographical background to research. Use notes and standard textbooks to research concepts and similar examples. Background on the location of the exercise is less important, but might be useful for exotic sites. The examples of issues analyses 1 and 2 on pages 71 and 82 show how important it is to demonstrate synopticity. (Allow a minimum of 2 hours.)

Individual/group work
Start this quite early on. Ask teachers for clarification of any on-going points. There might be information on the department intranet.

Explore opinions, check veracity (truth), assess the status of people and their emotional involvement in the issue, devise a conflict matrix, assess any reactions to schemes or sites (matrices are useful here) and highlight key points in statements for future use. (Allow 1 hour.)

For each option/scheme/site, work through the resources to synthesise details from *all* resources — this may involve map analysis. Make clear summaries. Look at all angles, such as positive and negative environmental or socioeconomic impacts, cost/benefit, and ranking from best to worst. Practise techniques for doing this and *keep results* for last-minute learning. (Allow 4 hours.)

Revision of notes and summaries
Do this the night before. Check you have used *all* resources fully. Revise factual details, for example synthesis of geography of the area and key features about options. Use diagrams to help you. (Allow 1 hour.)

How (or how not) to use your preparation time

- Reading the advanced information as many times as possible to learn the details and enable you to quote evidence ✔✔✔
- Annotating and making use of all the resources (e.g. highlighting text) ✔✔✔
- Gathering all the information about each option ✔✔✔

- Reading basic texts and notes on the geographical background ✔✔✔
- Working on the maps (OS and atlas) to gather as much information as possible ✔✔✔
- Discussing resources with friends to compare information ✔✔✔
- Analysing the opinions of people to check whether they are valid and to assess responses ✔✔✔
- Brainstorming what can be seen on photographs ✔✔✔
- Thinking about possible synoptic links suggested by the resources ✔✔✔
- Using a geography dictionary to check on terminology ✔✔
- Using the internet to obtain background information on the location ✔✔
- Reading round the topics in the advance information booklet and researching similar examples ✔✔
- Practising a range of techniques for options analysis etc. ✔✔
- Researching techniques in decision-making books ✔
- Planning answers to possible questions ✔
- Getting hold of further information about the site, possibly even visiting it ✘
- Trying to work out what the questions are ✘

Key

✔✔✔✔ Extremely important

✔✔✔ Important

✔✔ Quite important

✔ May be of some use

✘ Waste of time

Guidance on possible activities

Stage 1: sort out the resources

Sorting out the resources is probably best done during class sessions with your teacher, concentrating particularly on complex items such as a series of maps. You might use an overlay with a series of maps to 'sieve' out information. OS maps, digimaps and aerial photographs require intensive analysis.

Photographs are best *brainstormed* in class as different people will see different points. You can then put the ideas into a *geographical* description and annotate key points on the photographs.

There are various effective ways of annotating graphs, photos, diagrams and tables to summarise their *key* points. *Text* is surprisingly hard to analyse. Highlight the key points with marker pens and write shortened summary notes that synthesise key points. Look up key words in a geography dictionary.

Stage 2: researching the geographical background

Both the issues analyses examples in the Questions with Answer Guidelines section are for exotic places, with which you will probably not be familiar. Your centre may have invited an acquaintance, who has worked in or visited the location, to come and talk to the group about what it feels like to be there. This will give you a *sense of place*. Failing this, using websites may be useful, but don't spend more than an hour on this. It is more important to research the geographical *concept* and *issues* that the pre-release resources suggest might be relevant.

You need to define the terms and concepts and research similar examples, so that you can demonstrate *synopticity* via parallel examples. You have to be up-front in your answers to demonstrate linkages. Table 30 shows the common knowledge base in the specification, which you could draw on for synoptic issues analysis.

Table 30

Unit	Common ground
(1) Changing landforms and their management	• Processes and management • Sustainability
(2) Managing change in human environments	• Any land-based activities • Sustainability
(3) Environmental investigation	• Fieldwork techniques • Environmental impact • Monitoring strategies
(4) Global challenge	• Biodiversity • Changing weather and climate • Population and migration • Development economics • Globalisation and global shift
(5) Researching global futures	• Research techniques from secondary sources

Note: It is possible to use other data from both the physical and human divisions of Unit 5 (e.g. hazards or health) but the whole issues analysis exercise cannot be on Unit 5, because the unit is option-based. Table 31 shows how this applies to the two issues analysis exercises in the Questions and Answer Guidelines section of this guide.

Table 31

Issues Analysis 1 (Aqaba, Jordan)	Unit source	Issues Analysis 2 (Soufrière, St Lucia)	Unit source
• Theories of economic development — features of a middle income country • The states of development continuum	Unit 4	• World Heritage status • World Heritage danger list • Criteria for selection • Similar World Heritage sites studied — issues	Unit 4
• Strategies for development — industrialisation • Special Economic Zones (e.g. China) • Free ports and free trade — their value • Export processing zones — their purpose	Unit 4	• Small island Caribbean economics — issues related to this • Banana trade problems	Unit 4

Issues Analysis 1 (Aqaba, Jordan)	Unit source	Issues Analysis 2 (Soufrière, St Lucia)	Unit source
• Ways of measuring development GNP/GDP versus HDI (Human Development Index)	Unit 4	• Assessment of landscape quality • Ecological quality • Biological diversity • Heritage quality • Site conservation issues • How management plans work	Unit 4
• Advantages of middle income countries — assessment of resource base • Disadvantages of middle income countries — the Jordan situation (politics)	Unit 4	• National Park status — significance • Marine reserves — purpose and significance • Management strategies	Units 1 and 2
• Coral reef issues — value of biodiversity • Threats and damage to coral reefs	Unit 4	• Ecotourism — definition and operation • Geothermal areas — cause, value etc. • Conservation of urban heritage areas • What are the main issues here?	Unit 2
• Environmental impacts of development • Economic benefits of development • Issues associated with shanty town growth (Shalala)	Unit 2	• Issues of conservation versus exploitation	Unit 4
• Role of tourism in economic development — multiplier effects, leakage • Carrying capacity • Volatility of tourism in political circumstances (parallel examples)	Units 2 and 4	• Tourism industry — environmental, socio-economic and cultural impacts • Carrying capacity	Unit 2
• Sustainable management • Practical examples	All units	• Sustainable management • Practical examples • Features of SMMA summary	All units

Stage 3: decision making

Scour the resources to see if there are any possible schemes, sites or options that you might be called upon to analyse — in some issues analyses, there may be just one scheme. They will form the focus of the *decision-making element* in an issues analysis. You will be unlikely to be asked for a *full* evaluation of all options, schemes or sites. Given the breadth of the tasks and the limited time available, there is a wide range of possibilities. For example:

• You might have to choose the best option in terms of economic benefits, or the worst in terms of environmental damage.

- You might be asked to justify a rank order, or the results of a cost/benefit or environmental impact analysis, for each scheme.
- You might have to produce an annotated site map for a rejected option, or write a report on its limitations.

You should synthesise and summarise the information about the schemes, options or sites identified. At this stage, you could use the whole range of techniques developed during your A-level course. Table frameworks (use a computer spreadsheet or big wall charts) are useful tools to help you collate the details from each resource, ready for learning the night before the exam.

You could then attempt to produce a range of matrices, weighted scoring tables and ranking tables to cover a range of possible options.

A group could look at:
- strengths and weaknesses
- costs versus benefits
- advantages and disadvantages
- ranking
- environmental and socioeconomic impacts

Decision-making techniques or not?

You may have no choice — the required technique may be stated in the question. Otherwise, ask yourself:
- is there an obvious technique to use?
- is there a suggestion in the letter that a technique might be a good idea?
- It may be important for ranking but not applicable for evaluation. Lack of evidence is often a major problem. The technique must be fit for the purpose.
- Has it been pre-rehearsed? You may have carried out a technique that fits the question.
- Is the timing going well? Assembling evidence and using techniques takes time. Many candidates fail to finish as a result of choosing an inappropriate or over-complex technique.
- Could the technique be used for several questions? This would make the time spent more worthwhile.

Remember that you do not know the questions. Therefore, you have to develop a range of angles as part of your preparation. Learn the key features, so you are ready for action on exam day.

Stage 4: opinions analysis

A range of techniques for opinions analysis is suggested on page 56. Where opinions are listed, assess these to check whether they are justified, how they conflict with each other and whether they are for or against particular schemes or options.

Questions
with Answer
Guidelines

In this section of the guide, there are two issues analysis examples based on the topic areas outlined in the Synoptic Guidance section (Table 31 on pages 66–67 shows how they link to the specification.)

When answering at A2, you must be prepared to assemble your response in coherent prose. The issues analysis is unusual in that it should be written as a report with incorporated tables, matrices and annotated maps and diagrams. Be very careful in the use of bullet points as they must contain sufficient detail and coherent linkages.

Quality of written communication for whole report

Quality and cohesion of the whole report is worth up to 10 marks. This includes:
- following a sequence of enquiry
- using evidence effectively from the resources (quoting precise details)
- writing in a structured, logical format
- providing a report fit for purpose, i.e. using bullet points, tables and techniques effectively
- showing good use of geographical terminology
- demonstrating effective synopticity

Examiner's comments

Frameworks are supplied to help you plan your answers. These are indicated by the icon *e* and suggest how each answer should be approached. Guidelines for answers are provided after the resources.

The Aqaba special economic zone (ASEZ)

Once you have researched the necessary geographical background to the issue, study the resources provided and then carry out the following tasks.

(1) Jordan is classified as a middle-income country. Use evidence from the resources to justify this. (15 marks)

(2) Explain why countries such as Jordan see the designation of special zones as a key strategy for economic development. (10 marks)

(3) Evaluate the strengths and weaknesses of the Aqaba proposal. (15 marks)

(4) (a) Explain why it may be difficult to make **ASEZ** 'a beacon for sustainable development'. (10 marks)

  (b) What measures might you take to ensure that environmental sustainability is embedded in the **ASEZ** development? (10 marks)

Total: 60 marks

■ ■ ■

Framework for answering Question 1

- Define middle-income country.
- Look at Resource 1 and extract evidence to support the profile of a middle-income country.
- Obtain a range of evidence across both social and economic data.

Synopticity — refer to other middle-income countries you have studied. Consider the relative importance of economic development (GNP per capita) and the human development index (HDI). Discuss socioeconomic indicators in *relative* terms.

Framework for answering Question 2

- Explain why an SEZ could promote economic development (see Resource 2).
- Review the focus of the zone on tourism.
- Emphasise the importance of the position of Aqaba in Jordan's economy. Look at Resource 3.

Synopticity — Discuss the role of a raft of measures designed to support economic development. Refer to *any other* SEZ or freeport areas you have studied. You could relate to tiger economies and assess the likely impact on employment and the balance of payments at local and national levels of export processing zones.

1

Issues analysis

Framework for answering Question 3

e Many responses are possible here:
- Refer to the aims of ASEZ.
- Look at a balance of positive and negative environmental, economic, socio-cultural and political contexts at local, regional and national scales.
- Use a range of evidence.
- Resources 4, 5 and 6 are useful.

Synopticity — try to evaluate the proposal in the context of the potential conflict between development and conservation, or between local and national interests. Compare with other examples and discuss Aqaba's emphasis on tourism.

Framework for answering Question 4

e **(a)** Define sustainable development and look at the possible constraints of ASEZ. Resource 5 is a key resource.

Synopticity — try to look at multifaceted sustainable development (the suitability quadrant — ecofriendly, futurity, equity and community involvement) or the issues of trying to achieve environmental and economic sustainability. Use examples to support the identification of any difficulties (e.g. reef destruction and management).

(b) Safeguards and measures could be associated with:
- pollution control (reefs are a key issue)
- greening of the environment
- use of resources

Synopticity — provide examples of measures, such as reef management, tree planting and screening, use of solar energy and recycled water.

■ ■ ■

Resource 1: profile of Jordan

Official name:
Hashemite Kingdom of Jordan

Sharing borders with Iraq, Syria, Israel and Saudi Arabia, Jordan has just 26 km of coastline on the Gulf of Aqaba. Jordanian territory legally includes the West Bank of the Jordan river and East Jerusalem, but Israel has occupied these areas since 1967. Jordan ceded its claim to the West Bank to the PLO in 1988. Phosphates and tourism associated with important historical sites such as Petra are the mainstays of the economy.

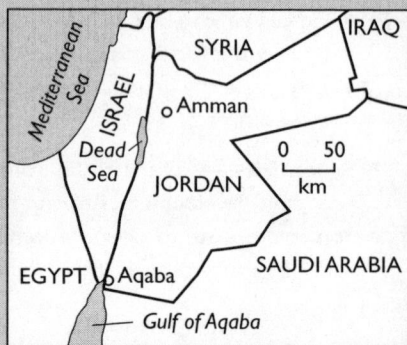

People

Population density: medium
Human dev. index: 94

67 per km^2

Arabic

Urban/rural population split

71%   29%

Jordan is a predominantly Muslim country drawn from Bedouin roots, with a Christian minority and a large Palestinian population. The monarchy's power base lies among the rural tribes, which also provide the backbone of the military. National identity is strong.

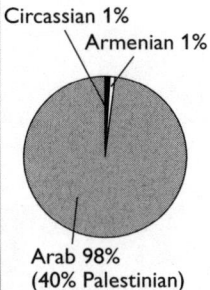

Circassian 1%
Armenian 1%

Arab 98%
(40% Palestinian)

Economics

Currency: Jordanian dinar

$= 0.71 Jordanian dinars

World GNP ranking	96th
GNP per capita	$1520
Balance of payments	$29m
Inflation (1985–1996 average)	4%
Unemployment	20%

Strengths:
Positive impact of 1994 peace treaty with Israel. Major exporter of phosphates. Skilled workforce. Recovery of tourist industry after 1991 Gulf crisis.

Weaknesses:
Reliant on imports to satisfy energy requirements. Unemployment owing to influx of Jordanians and Palestinians expelled from Kuwait.

Exports

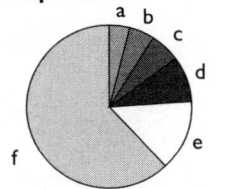

a — Syria 4%
b — Ethiopia 5%
c — UAE 6%
d — India 9%
e — Saudi Arabia 14%
f — Other 62%

Imports

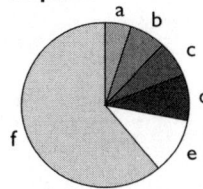

a — Japan 5%
b — UK 7%
c — Italy 7%
d — Germany 9%
e — USA 11%
f — Other 61%

Crime

Death penalty: used

No prison figures published

Down 65% from 1992–96

Jordan is largely peaceful. Crime levels are generally low, although theft in urban areas is rising.

Education

School leaving age: 15
Literacy: 87%

112 959 students

Men and women receive the same education. Jordanian teachers work all over the Middle East. 80% speak English.

Health

Welfare state health benefits available.

1 doctor per 625 people

Healthcare is subsidised by the government. Hospitals are found throughout the country.

World ranking

Position in world

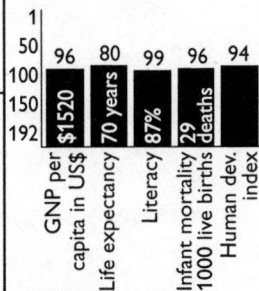

GNP per capita in US$	Life expectancy	Literacy	Infant mortality /1000 live births	Human dev. index
96	80	99	96	94
$1520	70 years	87%	29 deaths	

Climate

Summers are hot and dry, winters cool and wet. Areas below sea level are very hot in summer and warm in winter.

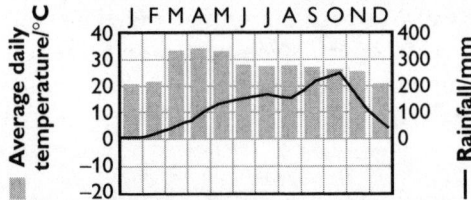

J F M A M J J A S O N D

Average daily temperature/°C
Rainfall/mm

Transportation

Adequate roads link main cities. A railroad links the port of Aqaba with the Syrian capital, Damascus.

6600 km

None

Queen Alia Int., Amman
2m passengers

293 km

None

7 ships
42 799 grt

Drive right-hand side

Tourism

Aqaba offers fine beaches, water sports and subaqua diving, while the ancient city of Petra attracts visitors interested in Roman remains. Amman is developing as a centre for Arabic culture and the arts.

1.1m visitors up 5% in 1995–97, down 25% in 2001–03

Main tourist arrivals

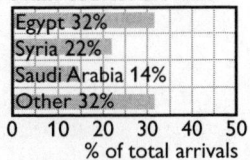

Egypt 32%
Syria 22%
Saudi Arabia 14%
Other 32%

0   10   20   30   40   50
% of total arrivals

Issues analysis

Resources	Spending	Environment
Oil deposits have been discovered. Phosphates, livestock and crops such as tomatoes, wheat, olives and vegetables are the main resources.	The wealthiest Jordanians are the entrepreneurs, bankers and engineers based in the capital, Amman. Poverty is relatively rare.	Conservation is a government priority. Rare animals are protected, and species that became extinct in the wild in the 1950s are being reintroduced into controlled environments.

Resources:
- 533 tonnes
- 23.3m chickens / 2m sheep / 795 000 goats
- 60 b/d (reserves 4 000 000 bbl)
- Oil, phosphates, potash

Spending:
- 50 cars per 1000 people
- 70 telephones per 1000 people

spending as % of GDP

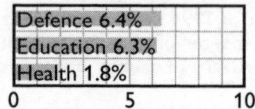

Defence 6.4%
Education 6.3%
Health 1.8%

0   5   10

Environment:
- 3%

Source: *DK World Desk Reference*, Dorling Kindersley

Resource 2: strategies for industrialisation

Most LEDCs and middle-income countries (MICs) see industrialisation as the key to economic development.

Developing countries employ one or more of the following strategies:
- **Exporting** indigenous commodities (resources and agricultural products); in particular, adding value to existing primary products or developing new ones.
- **Import-substituting industrialisation** — manufacturing products that would otherwise be imported. Protection strategies such as **tariffs** or **quotas** on imported goods are necessary to protect the fledgling industries.
- Creation of various types of **special zone**, for manufacturing or mixed development. In the 1980s, **export-processing** zones were favoured. Recently, special economic zones such as **freeports**, **free-trade** areas or **enterprise** zones have become the most common strategy. These zones rely on especially favourable investment and trade conditions.

Factors influencing the choice of strategy include:
- the resource endowment of the area — both physical (minerals, water, oil etc.) and human (e.g. the quality of the workforce), levels of education and foreign language proficiency)
- the size of the country, which affects the strength of the domestic market — a combination of the size of the population and its spending power
- the infrastructure
- the attitude of the national government towards enterprise
- the international context, including the political state of the area — instability affects investment

Resource 3: why invest in ASEZ?

An artist's impression of the new waterfront

- ASEZ is emerging as a major duty-free economic development location for tourism, recreational and environmental services, professional services and hi-tech and value-added industries.
- ASEZ is strategically located at the crossroads of three countries, at the northern tip of the Red Sea on the Gulf of Aqaba. It covers $375 \, km^2$ and extends along the whole of the Jordan coast from the borders of Saudi Arabia to Israel.
- ASEZ offers global business opportunities in an accessible and competitive location, with a high-quality lifestyle for would-be investors and residents.
- ASEZ has a good quality infrastructure and high-grade facilities. As well as a modern, full-service seaport and international airport, Aqaba has modern utility services — electricity, high-quality water, a waste water treatment plant, a fibre optic link around the globe and modern telecommunications services.
- Investors in ASEZ will benefit from Jordan's well-educated (almost 90% literacy rate) and relatively low-cost workforce. More than 17% have received higher education. There are training grants available for employment of Jordanian workers. Streamlined labour and migration procedures make ASEZ one of the most flexible labour markets in the Middle East.
- ASEZ is a duty-free environment ensuring special benefits, such as:
  – streamlined and simplified business registration
  – simplified foreign work permits and visas
  – no social services tax
  – exemption from most forms of sales tax
  – no animal, land and building taxes on utilised property
  – only 5% business income tax
  – exemption from customs duties on all imports to ASEZ (except cars)
  – no restriction on foreign currency or the repatriation of profits

A **Qualified Industrial Zone** within ASEZ allows manufactured products to benefit from duty-free access to the USA and EU.

Planned investment is 50% tourism, 30% service industry (e.g. business services), 13% heavy industry (e.g. port-based) and 7% hi-tech industrial development.

Resource 4: master plan and developments

The master plan has four visions. Aqaba will be:
- an oasis for commerce
- an up-market destination for tourism
- an incubator for modern technology (including hi-tech industries)
- a centre for environmental research and protection

The master plan is built around five development zones. While encouraging development, environmental sustainability is a key feature.

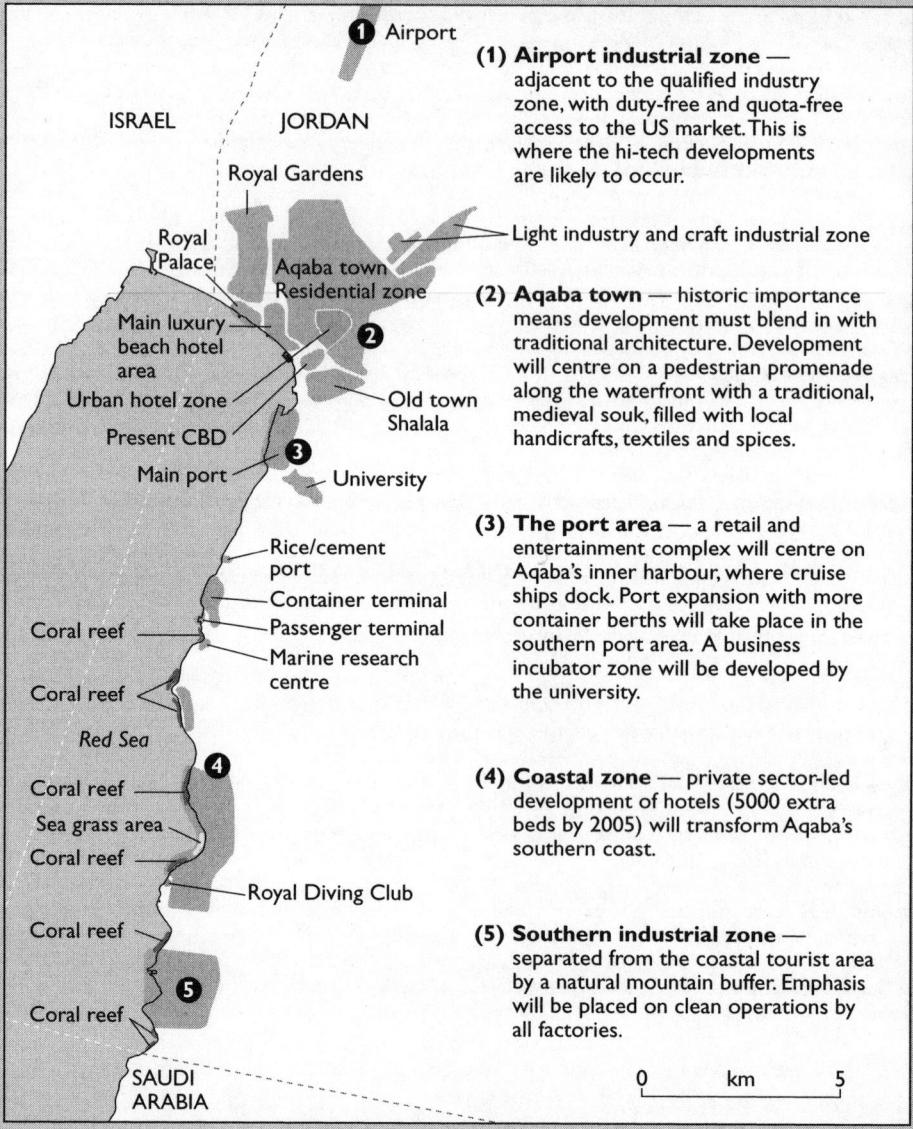

1 Airport

ISRAEL

JORDAN

Royal Gardens

Royal Palace

Aqaba town
Residential zone

Main luxury beach hotel area

Urban hotel zone

Present CBD

Main port

2

Old town
Shalala

3

University

Rice/cement port

Container terminal

Coral reef

Passenger terminal

Marine research centre

Coral reef

Red Sea

Coral reef

4

Sea grass area

Coral reef

Royal Diving Club

Coral reef

Coral reef

5

SAUDI
ARABIA

Light industry and craft industrial zone

(1) Airport industrial zone — adjacent to the qualified industry zone, with duty-free and quota-free access to the US market. This is where the hi-tech developments are likely to occur.

(2) Aqaba town — historic importance means development must blend in with traditional architecture. Development will centre on a pedestrian promenade along the waterfront with a traditional, medieval souk, filled with local handicrafts, textiles and spices.

(3) The port area — a retail and entertainment complex will centre on Aqaba's inner harbour, where cruise ships dock. Port expansion with more container berths will take place in the southern port area. A business incubator zone will be developed by the university.

(4) Coastal zone — private sector-led development of hotels (5000 extra beds by 2005) will transform Aqaba's southern coast.

(5) Southern industrial zone — separated from the coastal tourist area by a natural mountain buffer. Emphasis will be placed on clean operations by all factories.

0   km   5

Resource 5: The potential environmental impacts of ASEZ

Death knell for Jordan's coral reef paradise

AT THE end of the long dusty drive south on Jordan's Desert Highway can be found an army of ageing lorries laying siege to the Red Sea resort of Aqaba.

The thousands upon thousands of vehicles packed into the town's lorry park — a sea of rusting steel and iron stretching to the horizon on either side of the road — are evidence of the economic pressure that threatens to overwhelm the kingdom's 26 km coastline.

The port has become an increasingly threadbare lifeline as Jordan competes for international business. Trucks loaded with salt, phosphates and potash — the only major natural resources — rumble towards it, returning with food, livestock and oil. Hidden by the lorry park is Aqaba, a pale, low-rise reflection of Israel's popular resort of Eilat, whose towering hotels and apartment blocks on the other side of the narrow gulf seem designed to taunt its poor Arab neighbour.

Posters everywhere advertise the town's biggest tourist attraction: the brightly coloured fish that live among the fingers of coral.

The tourists are here to swim over these reefs, the most northerly in the world and, until now, some of the most pristine. In the sheltered waters of this small gulf, a delicate ecosystem has evolved, supporting more than 1000 species of fish and hundreds of types of soft and hard coral, many unique.

But the death knell is sounding for this underwater paradise. Akram Bederat, an instructor at the Royal Diving Centre, eight miles south of Aqaba, says the busy container port has wreaked havoc on the reef's flora. During the past two years, there has been a dramatic fall in the number of fish, even at great depths. 'The coral is a living thing full of bacteria and tiny organisms; stand on it and an area of one square metre is killed. It is like a crate with one bad tomato in it — all the rest are soon contaminated,' he says. 'And once an area is damaged it needs 20 years to repair itself. That can be a death sentence for the fish that live on the corals.'

He mostly blames the hundreds of tons of phosphates which are spilt each year during loading, showering the coral below with a fine dust that is suffocating the organisms. The export of Dead Sea salts has also raised the salinity of the coastal waters to dangerous levels.

Added to this, the gulf is being polluted by the increasing number of ships using the port. They flush out their tanks and, because the gulf waters take up to 3 years to make their way back into the Red Sea, the oil has longer to do its damage.

However, a new and bigger threat looms in the form of a free trade zone recently created to tempt international companies to a nascent industrial area close to the Saudi border. Petrochemical firms and fertiliser manufacturers are among the businesses queuing to do business. They are certain to add to the heavy burden on the gulf. Even before they arrive, the infrastructure to support the expansion is taking its toll.

Last year, four power cables and two telecommunication lines were laid between Egypt and Aqaba, cutting directly through the reefs.

A seawater-cooled power station in the industrial zone is trebling its capacity to cope with the new business, prompting concerns about what temperature changes in the water will do to the corals.

Bruce Pollock, a British scientist based in Aqaba with the Global Environment Facility, a World Bank project which is trying to limit damage to Israel and Jordan's coasts, says Jordan is aware of the dangers. 'It is a matter of balancing development with conservation, but for Jordan that's difficult because of the limited coastline.'

Environmentalists are equally worried by the regional authority's land-use plan, which was recently revised to take account of Aqaba's other boom industry — tourism. A 5-mile-long 'marine peace park' will be dominated by a holiday resort, golf course, tourist village, camping area and a Disney-style theme park.

Philip Reichel, a Dutch volunteer working for the Royal Ecological Diving Society, says: 'Along the 26 km coast there is not a metre of land they have not accounted for. How can the reefs survive when they will be damaged either by the port and free trade zone or by entrepreneurs exploiting them for the tourists?'

Because many of the reefs are close to the beach, holidaymakers can walk straight onto them to look at the fish, thereby killing swaths of coral. Mr Bederat says local bylaws mean visitors can be fined 20 000 dinar (£18 200) for damaging coral, but he has never heard of anyone being prosecuted. 'Warnings are needed on the beaches where they would be seen,' he says.

Source: the *Guardian*, 24 June 1998

Resource 6: some opinions about the Aqaba special economic zone

- **Resident of Shalala** — it is vital that we provide employment for people who live in very poor conditions in the old town of Shalala. Local unemployment is around 30%. Recently, in anticipation of ASEZ, squatter settlements have begun to grow.
- **Airport manager** — one problem holding us back at the moment is the Israel situation, which means that the airport has a very limited catchment. Aqaba International Airport has huge potential — unlike Eilat Airport, it has an extensive open site. The recent Middle East situation has hit tourism badly in this region.
- **Aqaba Tourist Development Board** — there is so much potential for tourism in Aqaba, such as reef tourism and day trips to the desert (Wadi Rum) and the World Heritage Site of Petra. It is vital that the waterfront area is cleaned up and developed into a thriving cruise ship port of call and the current CBD upgraded. There are limited modern shopping facilities. At present, ships only come twice a week.
- **Middle-class businessman** — while Aqaba has a reasonable area of shops, the local people would appreciate an enhanced shopping area. An air-conditioned shopping mall would be very enjoyable for us and the tourists. We are generally very much in favour of ASEZ, with its emphasis on business services and tourism.
- **Port manager** — the continued development of the port is vital for the whole of Jordan, as it is the only direct outlet to the sea.

Resource 7: current importance of Aqaba for reef tourism

In the picture below, note that the diving has little land-based support. The coral reef lies within and near the southern industrial zone — there is a phosphate boat in the background. Currently, the quality reefs are backed by an unattractive, unserviced land area.

Guidelines for answers to question 1

A middle-income country is one that has risen above LEDC status, both in terms of economic wealth and HDI.

Evidence that Jordan is a middle-income country includes the following:
- GNP per capita of $1520 (mid-ranking — 96th of 192 countries)
- HDI runs parallel with this — 94th of 192 countries
- There are a number of favourable indicators, including education. The literacy rate is high (87%) and there is a strong higher education element. There is no gender bias in education. There is one doctor per 625 people, which leads to a good life expectancy of 70 years (world ranking, 80th of 192). Poverty is rare. The economy is healthy, with a trading surplus. The population is 71% urbanised, which is indicative of strong economic development.

Guidelines for answers to question 2

Special economic zones are designed to support developing industry and services in a particular area. There is freedom from bureaucracy, tax regulations are favourable and grants are available to encourage investment (see Resource 3).

The focus on tourism, business services and high technology reflects the position of Aqaba as the only coastal area in Jordan (very high significance).

The SEZ will increase inward investment, improve the urban environment and provide a greater range of employment (30%) for a well-qualified workforce, playing to the strengths of Aqaba.

Guidelines for answers to question 3

	Strengths	Weaknesses
Environmental	• Should improve appearance of town centre, shantytowns, coastal promenade • Includes a centre for environmental research and protection • All very carefully zoned • Attractiveness of environment (climate) for all the listed activities	• Very fragile reef and desert environment • Issues associated with port/industrial development and tourist development — essentially direct damage, siltation and pollution
Economic	• Should bring numerous jobs to improve on 30% unemployment • Inward investment in up-market hotels and facilities should add value (e.g. a cruise destination would bring foreign exchange) • Diversification will be useful, producing a broad-based economy • Trade facilities will be strengthened	• Perhaps an over-concentration on tourism — a highly volatile industry • Could encourage economic leakage as imports are made for tourism
Social	• Improved facilities, such as malls and restaurants, for local people, many of whom will be upwardly mobile • Strong support for the scheme (see Resource 6) • May lead to clean-up of shantytown areas	• High-powered jobs may go to non-Jordanians, who may find a glass ceiling • Exposure to Western culture may cause conflict in a Muslim society
Political	• Will strengthen Aqaba's role within Jordan	• Conflict in Israel restricts development (airport and volatile tourism industry)
General	• Broad range of mainly high-quality, clean developments	• Perhaps overly dependent on tourism

Guidelines for answers to question 4

(a) The aim is to provide a more sustainable economy for Aqaba and to improve the lives of people through steady employment. It seems to be a top-down scheme, but well supported by local people. The very poorest (who have been 'sent back' as a result of the Gulf War) may lack education for the range of quality jobs on offer, so creating an underclass. There are also concerns over how ecofriendly the project is — in spite of environmental research being at the centre of focus. Environmental concerns can result from development. There are also issues of increased use of land, resources and, above all, water in a desert area.

(b) Measures might concentrate on:
- avoiding pollution of oceans and reef area by planting trees to cut siltation, legislation and insistence on clean technology to cut pollution and education of reef users
- avoiding increased land-based pollution by ensuring clean-up of waste land and litter, greening of the environment and landscaping of new facilities
- using resources effectively. Water is an issue (recycling and desalination plants). Climate statistics suggest solar power could be useful
- ensuring adequate public transport of workers and tourists between zones and formulating a green travel plan to reduce pollution emissions from cars
- involving local people in management, such as special training and employment grants to ensure improvement in the poor areas of Shalala

Issues analysis **2**

World Heritage status for Soufrière?

Brief

For the purpose of this exercise, you are working as an intern for the Soufrière Development Programme. The organisation promotes community-developed sustainable tourism, which is vital for the employment of the people of Soufrière (the current unemployment level is around 35% and rising).

Your tasks include:
- advising on St Lucia's bid for Soufrière to become a World Heritage site. Advice is needed on whether to go for natural-site or mixed-site status and on which of the four major tourist sites to include
- using the whole capacity of the immediate surroundings to bring tourists to Soufrière and satisfy their holiday interests
- evaluating the likely economic and environmental impact of the following schemes:
  – the development plan for Sulphur Springs
  – the plans for the development of Soufrière town
  – the introduction of ecotourism projects in the Pitons
- undertaking a week's work experience at Soufrière Marine Management Area (SMMA) and researching this unique experience in sustainable management

The further development of tourism is vital for Soufrière.

(1) Outline the advantages and disadvantages of achieving World Heritage status. You should refer to other sites you have studied as well as to St Lucia. (10 marks)

(2) Write a report to support Soufrière's application for World Heritage status. You should advise on the type of status and size of the area, fully justifying which of the four key areas to include. (20 marks)

(3) Assess the likely environmental and socioeconomic impact of the following plans:
  (a) development of Sulphur Springs
  (b) improvements to Soufrière town
  (c) ecotourism projects on Gros Piton (20 marks)

(4) Explain how SMMA fulfils the requirements of a sustainable management scheme. (10 marks)

Total: 60 marks

■ ■ ■

Framework for answering Question 1

e Read Resource 2 very carefully and think about:
- why countries apply for World Heritage status
- what the environmental and economic consequences might be

Synopticity — in Unit 4 you are required to study two World Heritage areas. Use examples from these (e.g. the Galapagos Islands) to inform your answer. Advantages include prestige and enhanced conservation. Disadvantages include the carrying capacity being overwhelmed by tourists.

Framework for answering Question 2

e • Status — natural, cultural or mixed
- Size of area — which of four key zones should be included (see Resource 1)? Relate to the criteria in Resource 2.

Synopticity — evaluation of landscape, natural phenomena and levels of biodiversity. Use knowledge of reefs from Unit 1 and Unit 4 for biodiversity. Assess levels of conservation currently on offer — National Park status and the marine conservation area of Soufrière reef.

Framework for answering Question 3

e Remember to look at:
- both advantages and disadvantages
- a balance of environmental and socioeconomic impacts

Synopticity — you will have had an opportunity to develop impact matrices. Use information from Unit 2 to look at issues of honeypot site management, urban heritage tourism management and the role of ecotourism in the Piton area. Refer to similar examples you have studied.

Framework for answering Question 4

e • Define sustainability
- Key requirements include futurity, ecofriendliness, community participation and equity

Synopticity — use examples of other sustainable management schemes you have studied. Think about the conservation of marine resources (Unit 4), the value of ecosystems (goods and services) and the issues involved in creating marine and fishing areas. Assess the ecofriendliness of the scheme, the degree of community participation and conflict resolution between groups, especially fishermen. Remember the LEDC context.

■ ■ ■

**Issues
analysis**

Resource 1: introduction to St Lucia

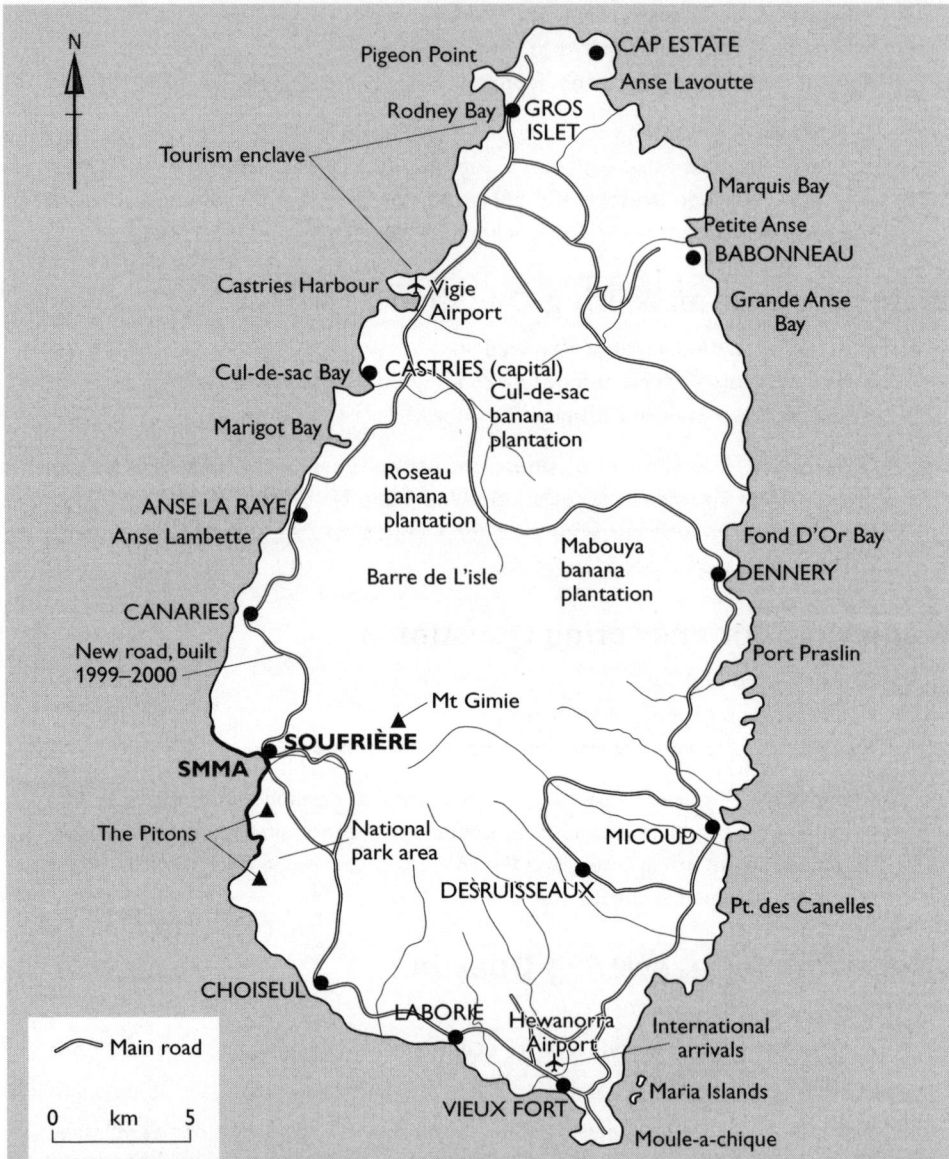

N

Pigeon Point

CAP ESTATE

Anse Lavoutte

Rodney Bay

GROS
ISLET

Tourism enclave

Marquis Bay

Petite Anse

BABONNEAU

Castries Harbour

Vigie
Airport

Grande Anse
Bay

Cul-de-sac Bay

CASTRIES (capital)

Cul-de-sac
banana
plantation

Marigot Bay

Roseau
banana
plantation

ANSE LA RAYE

Anse Lambette

Mabouya
banana
plantation

Fond D'Or Bay

Barre de L'isle

DENNERY

CANARIES

Port Praslin

New road, built
1999–2000

Mt Gimie

SOUFRIÈRE

SMMA

The Pitons

National
park area

MICOUD

DESRUISSEAUX

Pt. des Canelles

CHOISEUL

LABORIE

Hewanorra
Airport

International
arrivals

Main road

Maria Islands

0   km   5

VIEUX FORT

Moule-a-chique

St Lucia is a small Caribbean Island with a population of 150 000. It is part of the Windward Islands. It is a middle-income country, with a GNP per capita of US$3510, and its HDI rank is 81/192. However, its economy is typical of a small island developing economy, with a trade deficit of $80 million. Its economy is vulnerable to outside influences, being very dependent on bananas (facing the problem of removal of preferential trading with EU) and tourism (a volatile industry).

St Lucia has a rich natural, cultural and historical diversity and attracts nearly 500 000 visitors each year. The wide range of microclimatic zones on the island provides the setting for many habitat types and a wealth of biodiversity. Until the 1990s, the economy of the island was largely based on agricultural exports — mainly bananas. However, in recent years, tourism has become the dominant economic activity, with Soufrière the main focus. The main hotel belt is an enclave in the northwest of St Lucia.

Resource 2: what is a World Heritage area?

Under the terms of the World Heritage Convention, a World Heritage list has been established of properties of outstanding universal value, which form part of the exceptional natural and/or cultural heritage of the world.

Heritage sites may be:
- cultural — masterpiece buildings (e.g. the Taj Mahal) or whole cities
- natural — scenery, geology or biodiversity (e.g. the Grand Canyon)
- mixed — combining ruins and scenery (e.g. Uluru, Machu Picchu)
- in danger — from natural disasters or human actions, such as war or pollution; assistance is provided to restore site

In 2002, there were over 700 sites, the distribution of which is summarised below:

Area	Cultural	Natural	Mixed	In danger	Comment
Africa	57	32	3	16	25% of natural sites; highest number in danger
Asia	132	28	6	9	China leads list (28); includes rainforest
Australasia	0	13	5	0	Cluster concept; very good conservation ethic
Europe	289	23	6	3	Impact of long history
North America	48	29	1	3	Culture mainly in Mexico; many National Parks in USA and Canada
South America	37	19	2	2	Many archaeological sites; short on rainforest sites
World	563	144	23	33	Increasingly varied coverage, including Victorian sites (e.g. Ironbridge Gorge) and geological sites (e.g. the Dorset Coast)

There is a complex selection process before a site is placed on the World Heritage list.

2

The World Heritage Committee receives a detailed report form each applicant. To be included on the list for natural heritage, the location must fulfil *at least* one of the following criteria:

- an outstanding example of a site representing major stages of the earth's geological history (e.g. Dinosaur National Park, USA)
- an outstanding example of a site representing significant on-going ecological and biological processes in the evolution and development of plants and animals (e.g. the Galapagos Islands)
- contain superlative natural phenomena or areas of exceptional natural beauty and importance (e.g. the Grand Canyon)
- contain the most important and significant natural habitats for conservation of biological diversity — particularly rare and threatened species (e.g. the Great Barrier Reef)

A site is selected according to the following criteria:

- value — its significance, authenticity and representativeness
- integrity — an area large enough to ensure its preservation
- state of protection — what is being done to protect it? What signs of degradation are there? What likely pressures are there on the site?

Resource 3: Soufrière development programme

The Pitons are the gateway to the Caribbean. They provide a variety of opportunities for recreation, including nature appreciation, hiking, wind surfing, sailing, snorkelling and scuba diving.

The goal of the Soufrière development programme is to promote coordinated socioeconomic development, creating amenities and infrastructure in a socially acceptable, environmentally sound and economically beneficial way.

Tourism is perceived as a key element of the development strategy because of the area's natural and historic attractions and because that sector has potential for employment and investment opportunities.

The plans call for intensive development of small-scale, locally owned lodgings, shops and associated facilities in Soufrière town, coupled with the recreational and scenic attractions in the vicinity, including the Pitons National Park.

Soufrière is unique, with natural and historic features found nowhere else in the world. As all Caribbean countries seek to develop their tourism sector, Soufrière has a chance to become a tourist destination capable of competing with any other.

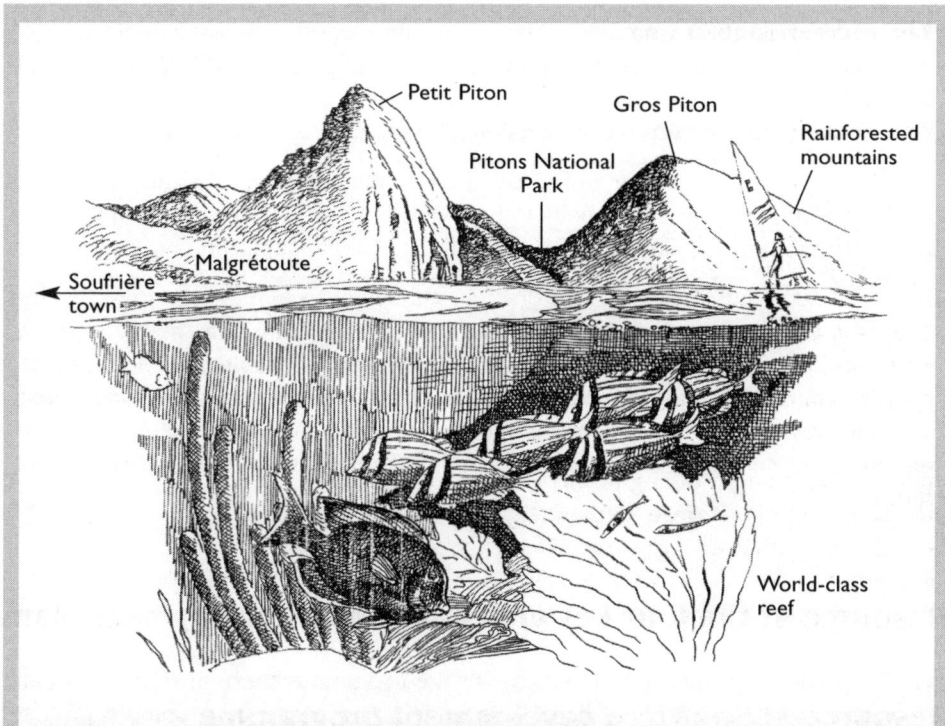

Among the many resources that must be preserved and developed for the benefit of visitors are:

- outstanding landscapes
- sulphur springs
- beaches
- rainforests
- Soufrière town's architectural heritage
- coral reefs
- historical estates
- archaeological sites
- wildlife

Tourism and Soufrière

Due to the historical circumstances, Soufrière has a unique potential for socially integrated and locally owned tourism development. The natural beauty of Soufrière and its surroundings is augmented by an exceptional development opportunity at the waterfront of Soufrière Bay. There are potential links to the fishing, agriculture, agro-processing and handicraft sectors.

The image of Soufrière is integral with its natural setting. The Pitons to the south, the backdrop of rainforested mountains to the east and the steep hills and cliffs to the north provide a dramatic setting for this waterfront town's development as a tourist destination.

2

**Issues
analysis**

The recreation dilemma

The Soufrière area has limited beach facilities. The closest are at Malgrétoute and between the Pitons. The availability of those beaches for locally owned tourism development and resident recreational use is essential.

To bring tourists to Soufrière and satisfy their holiday interests, the full capacity of the immediate surroundings needs to be utilised. Short-stay tourists require ease of access and thus proximity to recreational facilities. Beaches and parks adjacent to Soufrière are a prime requirement.

Tourist growth in Soufrière is a delicate matter. To transform a town into a successful tourist haven is an ambitious challenge. The ability of Soufrière to provide a beautiful setting, a variety of recreational locations, clean air and water, and a healthy marine life becomes not only an ecological and planning issue, but also an economic necessity. We have to inspire the people of Soufrière to grasp the future.

Resource 4: the four key areas and their development plans

(1) The Sulphur Springs and Diamond Falls are in a unique thermal area. 25% of all visitors to St Lucia come here. It is undergoing a site development project involving improving facilities and community training programmes.

(2) Soufrière town was the first capital of St Lucia. It has many historic buildings surrounded by original sugar estates. There are two large hotels and several small up-market ones, but the number of residential places is limited. Many visitors, particularly those from cruise ships, are day-trippers.

(3) Gros Piton and Petit Piton are St Lucia's national landmarks and have National Park status. Currently, only around 1000 people climb the Pitons each year. There is potential for ecotourism, such as horse-riding.

(4) The SMMA includes an outstanding coral reef which is managed as a sustainable development initiative. The reef is visited by 21000 snorkellers and 12000 individual dives are made each year.

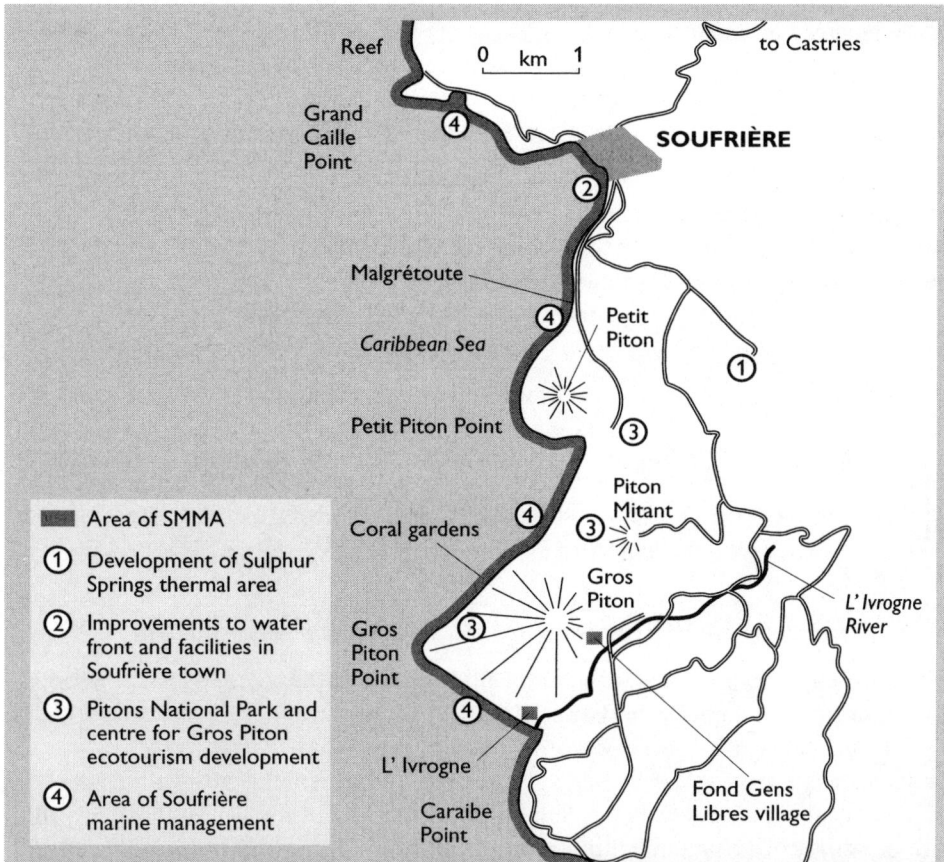

Area of SMMA

1. Development of Sulphur Springs thermal area
2. Improvements to water front and facilities in Soufrière town
3. Pitons National Park and centre for Gros Piton ecotourism development
4. Area of Soufrière marine management

Upgrading the Sulphur Springs

The geological formation called Sulphur Springs is located in Soufrière. The 25 acre site is a major tourist destination in St Lucia, drawing upwards of 120 000 visitors a year. The Sulphur Springs site is a unique phenomenon. It is widely publicised, frequently mentioned in conversation and print and is an integral part of a visit to the island. Sulphur Springs and the Soufrière area are the primary day excursion destination for cruise ship passengers and stay-over tourists.

The large number of visitors to the springs attests to the site's natural drawing power rather than to its development for visitor enjoyment, as amenities are few and inconvenient. However, the Sulphur Springs site has physical and economic potential. Physical development of the site, along with signage and trail development, would offer the visitor a memorable experience. In terms of economic development, it represents an unserved, almost captive market of visitors who would pay more than the present entrance fee to visit an interpretation centre and the site and who would purchase gifts, souvenirs and snacks.

2

Issues analysis

Development project

The development project will support environmentally appropriate development of a unique, natural phenomenon as a tourist attraction and provide opportunities for businesses catering for tourists. Thus, employment and income in the tourism and small business sectors will be increased.

The project as proposed includes:
- upgrading an existing road to become the new access road to the site
- relocation of displaced building rubble and general site clean-up
- construction of a suitable entrance and interpretation centre
- provision of site improvements, such as trails, viewing areas and public conveniences
- provision of utilities
- preparation of appropriate publicity material
- technical assistance and training in the fields of management and marketing, including creation of a strategic business plan for operation of the facility — a key element of the development's future sustainability

One key assumption is that the project will result in a sufficient increase in tourist expenditure at Sulphur Springs to make the investment self-sustaining.

Development impacts

There will be three groups of **beneficiaries**:
- transport interests — tour operators, taxi drivers and car rental agencies
- local residents — they will benefit from increased tourist visits to Soufrière, increased staffing levels at Sulphur Springs and through establishing and operating businesses catering to tourists
- national tourism interests — an improved visitor experience at Sulphur Springs could increase overall tourism in St Lucia

The project's **sustainability** will depend upon revenue generation and the management and operation of Sulphur Springs as a commercial business whereby excess revenue is reinvested in product improvement.

Preserving Soufrière Town's cultural heritage

Soufrière welcomes an estimated 120 000 tourists annually, drawn to the area by the natural environment, the history and the architectural heritage.

The Soufrière Foundation plans to develop these features and other cultural landmarks in a way that conserves them. This should result in an expansion of tourism, higher employment, increased income and net foreign exchange earnings.

The development of the waterfront/mall/square area will:
- provide a major attraction in terms of urban landscape and a focus for tourist activities

- facilitate the arrival and departure of tourists from cruise ships, boats and buses
- improve the area for social gatherings and business activities for the townspeople, creating a market place, conversational area and pedestrian-only promenade
- showcase the historic character of the town

The **Heritage area** of the old town is to be preserved and a historic architectural walk is to be introduced:
- to preserve and revitalise the buildings of interest
- to recreate the history and culture of the town

The **Soufrière Heritage Centre** has two objectives:
- to create a specific point of interest for visitors, and thus enhance Soufrière's tourism attractiveness
- to explain and conserve the cultural history of Soufrière

The Heritage Centre is intended to give visitors a view of the region's geological and cultural history, including the use made of natural resources by successive cultures of the region. The importance of French culture in the history of Soufrière and St Lucia will be highlighted, with the old prison, formerly a hospital, being restored. The area includes the Anglican Church, another important historical building. Both the prison and the church date from the early 1800s. Together they will form the focus of the proposed walk.

Additionally, the Soufrière Development Programme aims to develop facilities and improve the quality of souvenirs available in the market.

Many new rooms, in high-quality hotels, villas and guest-houses, are expected to become available in the region. The plan features a complex of small, *locally owned* hotels, arcades, guest-houses, restaurants, shops and museums on the waterfront. This would be coupled with the new pier and boardwalk marina complex and other waterfront improvements. Infrastructure works include a new sewerage system, improved roads, water supply and garbage collection.

The scheme is ambitious but feasible. Many of the public works projects are either underway or at an advanced planning stage. For instance, the pier and boardwalk marina complex have been completed. St Lucian entrepreneurs have shown interest in developing hotels, restaurants and shops. Financial requirements are within the capability of local financial institutions.

The scheme will transform Soufrière and its surrounding region. Increased income from stay-over tourists, day tours and cruise ships could revolutionise the economy. However, the key to the success of the project is the establishment of a principal recreational and marketing attraction in the immediate area.

Ecotourism for Gros Piton

The Pitons form the gateway to St Lucia and share with Soufrière a symbiotic relationship that will serve as a catalyst for the Caribbean's first socially acceptable, environmentally sound and economically beneficial tourism development.

The Pitons National Park is a Caribbean heritage site of international importance. It is the principal recreational attraction that will create the marketing force necessary to unleash the unique potential in Soufrière for the establishment of independent, locally owned tourism-related businesses and facilities, thereby maximising the economic benefits for the community.

The Pitons National Park

Features include:

- beach and marine facilities — shelters, picnic areas, docks
- visitor centre — rehabilitation of a historic rum distillery and sugar mill
- orientation village — dining, gift sales, cultural and historical entertainment
- botanic gardens — trails and signage for walking tours of tropical vegetation
- fruit and spice plantations — trails and signage for tours of tropical agriculture
- outstanding landscapes and natural ecological regions — marine, arid, deciduous rainforest and elfin or dwarf woodland

The ecotourism proposal

The proposal has several strands:

- self-guided and guided walks around Fond Gens Libres, to look at vegetation, nature and history and including interpretative lookout points and picnic facilities. The nature trails would include coastal hikes, mountain and ridge walks and easy strolls
- sea kayaking routes from Soufrière around the Pitons to L'Ivrogne, where camping facilities will be developed
- research into mountain biking trails and pony trekking as well as sites for hang gliding

People will experience:

- **the Pitons** — remnants of a large volcano that stood over the Soufrière region 40 000 years ago. Gros Piton is higher and more massive than Petit Piton but is easier to climb.
- **vegetation** — ranging from arid deciduous woodland containing cactus and thorn scrub on the coast, to a broad middle-zone of rainforest, an upper mountain zone of elfin woodland with windswept dwarf forest at the summit
- **birds** — Gros Piton provides a superior habitat for nesting birds and good shelter and forage for migratory species. Birds are abundant, especially in the cloud forest at the summit, where the shy St Lucia blackfinch can be found. The mangrove cuckoo is common near Fond Gens Libres. Numerous other species occur in the area.

- **legend** — the Pitons had religious significance to the island's original inhabitants, the Caribs. Gros Piton was referred to as the god Yokahu, who represented fire, thunder and food, and slept in the Sulphur Springs. Petit Piton was referred to as the goddess Atabeyra, who represented fertility, moving water, the tides and the moon
- **brigand history** — during the slave rebellion of 1748, this area was a hiding place for runaway slaves. Later, Gros Piton became a secure haven for black freedom fighters or 'brigands'. Brigand sites are numerous and include caves, tunnels, rock shelters, camps, signal stations, lookouts and landing sites

Resource 5: the Soufrière Marine Management Area (SMMA)

The biodiversity of the reefs around Soufrière is exceptional. The reefs were under increasing pressure from competing human activities, including direct effects of divers and other tourists, and from overfishing. Indirect pressures included siltation (deforestation of watersheds) and pollution from increased land-based development. The result was the creation of marine reserve areas (where fishermen were not allowed) and fishing priority areas (where yachts were not allowed to anchor). This created conflict between reef users and the marine areas became just 'paper parks'. (See **www.smma.org.lc**)

Between 1992 and 2002, SMMA was developed to solve these problems, conserve the reefs and allow equitable usage and bottom-up involvement.

Phase 1: 1992 to 1994
Key issues were identified:
- controlling land-based pollution
- delimiting reserves
- finding appropriate yacht anchorages
- protecting fishermen's livelihoods

Phase 2: 1993 to 1995
A period of conflict resolution and participatory planning was followed by the establishment of SMMA in June 1995. After consultation with user groups (stakeholders), a zoning strategy was developed. The stakeholders formed the management committee.

Phase 3: 1996 to 1997
This was initially a period of problems. As the zones were established, conflicts took place because fishermen were not allowed to set traps in the marine reserves. The new yacht anchorage was a local crime spot and not a success. Gradually, SMMA developed the necessary management structures for monitoring infringements of zoning.

Phase 4: 1998 to 2002

Successes included:

- improved quality of the reef resource and larger catches for the fishermen
- gradual empowerment of the community with participation of all major stake-holders
- economic viability — from fees, so that the education mission could be funded and rangers trained
- international recognition of both the process and the product

Guidelines for answers to Question 1

World Heritage status involves recognition of the quality of natural or cultural features. It is awarded to international standards.

Advantages	Disadvantages
• Could lead to greater protection and conservation of the site	• Excess popularity, which can put the site under threat
• Funds could be generated for further conservation, interpretation and management	• Issues include carrying capacity, erosion and litter
• World Heritage sites become magnets for international tourism, which can generate funding	• Example of Galapagos Islands — excessive tourism and overfishing

Guidelines for answers to question 2

If all four areas were included, a site of about 7–8 square miles would be formed, which would be costly to administer. The varied nature of the natural and cultural attractions would make a mixed-site bid necessary (Soufrière cultural area). On the other hand, a large site has enough area to ensure integrity. The components fit together, providing diverse landscapes.

- Sulphur Springs is a geothermal area with high drawing power. It is already visited by 120 000 people annually. Nevertheless, the site is poorly developed (little value added), so World Heritage status would help here. Currently, the state of protection is insufficient.
- Soufrière town has important cultural significance. Its nineteenth century French heritage is very significant in Caribbean terms. Currently, the town is visited on cruise ship days, but is in need of refurbishment. World Heritage status could lead to improved quality of the environment and the development of employment opportunities — there is currently 20% unemployment in Soufrière.
- The Piton area is of worldwide renown and already has National Park status; attractions include geology, vegetation, wildlife and legends/history. Currently it is under-visited. Ecotourism projects would offer an exciting and sustainable future.

- SMMA is an international-class coral reef. It is currently part of a sustainable development. The area would enhance the whole bid, as it would provide a coastal dimension.

The combination of all four sites could release some international pressure on them by providing a greater area for the visitors to explore, on cruise ship days and during longer stays.

Guidelines for answers to question 3

All the schemes are sustainable, which will support both local people and the environment.

(a) The Sulphur Springs scheme could improve the area, which is currently afflicted by poor quality guides, few facilities and low standards of interpretative materials. However, the visual impact of new facilities could be negative. There will be fragile environment issues if the carrying capacity increases beyond 120 000. Socioeconomic impacts are favourable. There is a need to provide jobs in Soufrière and there is the potential multiplier effect of long-stay tourism.

(b) Improvements to Soufrière town would definitely improve the environment quality — many schemes/grants are available. Culturally, education of tourists should support this. Economically, a multiplier effect should occur, with a range of opportunities in employment.

(c) The scheme is environmentally friendly (use of ecotourism) and should cause few environmental problems, although there could be some issues, such as trampling, associated with horse-riding. Key jobs in the village of Fond Gens Libres could also be a bonus.

Guidelines for answers to question 4

SMMA is an ecofriendly scheme to safeguard the environment of the reef for future generations, while providing a livelihood for local fishermen. Marine protection areas can enhance the quality of fish stocks. The unique development via stakeholder groups is bottom-up and aims to support some of the poorest members of the community. SMMA aims to achieve environmental sustainability, combined with economic viability of the Soufrière community. It provides a range of employment opportunities. Fees from dives and similar tourist activities will make it financially self-sustaining.